P9-DNU-831
WE-THINK
CHARLES LEADBEATER

WE-THINK

'*We-Think* is a riveting guide to a new world in which a whole series of core assumptions are being overturned by innovations on the web. Leadbeater draws a series of remarkable conclusions' Matthew D'Ancona, *Spectator*

'An important book, even for sceptics like me. *We-Think* is inspiring in its analysis, I urge you to read it' Andrew Keen, *Independent*

'I was gripped. The book's theme is as big and as bold as it gets … should be compulsory reading for all who seek to understand the driving force of this century' *Management Today*

'Helps readers to frame some of the important questions for the coming decade' *Director*

CHARLES LEADBEATER is one of the world's leading authorities on innovation and creativity in organisations. He worked for the *Financial Times* for ten years and was ranked by Accenture as one of the top management thinkers in the world. The *New York Times* described his idea of the Pro Am revolution as one of the most influential of the decade.

WE-THINK

Charles Leadbeater

(and 257 other people)

Illustrations by

Debbie Powell

PROFILE BOOKS

This second edition published in 2009

First published in Great Britain in 2008 by
PROFILE BOOKS LTD
3A Exmouth House
Pine Street
Exmouth Market
London EC1R 0JH
www.profilebooks.com

10 9 8 7 6 5 4 3 2 1

Typeset in Palatino by MacGuru Ltd
info@macguru.org.uk

Printed in the UK by CPI Bookmarque, Croydon, CR0 4TD

A CIP catalogue record for this book is available from the British Library.

ISBN 978 1 86197 837 0

For Harry and Ned

CONTENTS

PREFACE

I have tried, imperfectly, to write *We-Think* in the spirit of the argument, openly and collaboratively. The book draws on the ideas of many other people, who are noted in the acknowledgements. But about half-way through the writing it dawned on me that it would be odd to write about the growth of collaborative creativity in the traditional way: the writer at his desk, isolated from the world, alone with his thoughts. With the support of my publisher, Profile, I posted an early draft on my website so people could download it, print it, read it and comment on it. They could also go to a wiki version to change the text and distribute it to their friends and colleagues. And so I just let it go, bouncing along the links that make up the web like a skimming stone.

At first sight this is a very odd thing for a writer to do, on at least two counts. First, as several people pointed out, if I gave away the draft for free, would people want to buy the finished book? My hunch, confirmed by other experiments of this kind, is that sales will not suffer. The more the draft is downloaded, the more talked-about it will be and the more likely people are to buy the final work – all the more so as it includes Debbie Powell's great illustrations. The version you are reading today differs markedly from the first draft put on the web. Secondly, why wash my dirty linen in public? Showing a draft to anyone induces in me a deep insecurity and anxiety. There are bound to be errors, omissions, mistakes. That is why normally I show a draft only to my wife. Why on earth make it available to lots of people I do not know?

Since I put that early draft online in October 2006, the material has been downloaded on average 35 times a day; about 150 comments have been posted on the site about the text; it has been mentioned on more than 250 blogs; I have received about 200 emails from people wanting to point me in the direction of useful information; and by late 2007 a Google search for the book title and my name came back with 65,600 hits.

Did this little experiment in collaborative creativity work? Well, no one was horrible. There was neither vandalism nor abuse. Some of my early callers were pretty sceptical. The first post on the site, from an ardent Irish blogger, basically said, 'Who the hell do you think you are? I've been blogging for years – what do *you* know about it?' One respondent said the idea was 'codswallop' and another that it was 'stale'. Some people wondered whether it was just a clever wheeze to get other people to write a book for me and so to make money out of their voluntary contributions. Later, another respondent suggested I put a 'Donate here' button on the text to make sure I got paid: he warned me that lots of fake books were circulating on the Internet and suggested that if I were not careful someone would run away with my ideas.

Everyone was very forgiving of my poor spelling and grammar. Many people suggested improvements in how it could be published online, to make it easier for people to engage with. The system I had designed to enable people to make comments on the text proved far too cumbersome. Lilly Evans suggested that the book should be designed for the iPod generation so the contents could be shuffled about. Dave Pawson pointed me to some open-source software that would allow people to annotate the text. Lots of people directed me to material, links and examples they thought would amplify the argument, including Matt Hanson, who alerted me to his open-source film-making project, A Swarm of Angels, which involves hundreds of people; and Sandra from Vancouver, who told me about user-generated comic characters on mobile phones. Others let me know that they were taking up the ideas even before the book was finished. Michiel Schwarz and some friends in the Netherlands started a think-tank called The Beach in honour of some of the ideas that were in Chapter 1 when the draft appeared on the web and when I met them they introduced me to the idea that 'you are what you share' – which I then took as the title for the opening chapter. A television producer from Samsung in South Korea emailed to say he had been reading the text and wanted to interview me for the company's internal television channel. Paul Mark said he was trying to promote participative approaches to rural development in India. Virtually everyone was very supportive, which, of course, is very encouraging.

A number of people took the time to make really detailed comments, which were often challenging and improved the book no end. Miranda Mowbray wrote me a long and very helpful email on a Saturday morning from Bristol which among other things pushed me to think about the resurgence of folk culture. Nigel Eccles, a management consultant, posted a useful 10-point critique which tempered my overenthusiasm. Jeremy Silver corrected what I had said about his company Sibelius. Tim Sullivan wrote from New York to alert me to Scott Page's work on diversity and creativity which influenced my thinking a great deal. The most persistent, however, was Heiko Spallek, whom I never either met or talked to. Heiko is an assistant professor of dentistry at the

University of Pittsburgh and he read every last word, pointed out many errors and omissions, alerted me to interesting stuff he had seen and then at the end came up with his own conclusions for the book, which are available online with the text.

What my own feeble, partially successful experiment with We-Think has taught me is this. It is quite time-consuming if you want to do it properly; it takes daily attention. The conversation does not just take place where you happen to have set out your stall; it takes place in lots of different places, on blogs and sites all over the web. If you trust people and throw things open, they will respond. But you have to live with a degree of transparency that might be uncomfortable: in my case everyone can see I cannot spell and have no idea what a comma is for. People will contribute when they feel motivated to do so. Most people who contributed were not that interested in contributing to my book. Why should they be? They wanted to connect to ideas *they* found interesting, for *them*, in *their* lives. The trick is to gather that self-interest for mutually beneficial ends. I cannot imagine writing another non-fiction book in any other way. Next time I would start putting material online much earlier and make it a lot easier for people to make suggestions, so people have more opportunity to talk about it, shape it and contribute to it.

But it would be misleading if I claimed that this book were mainly the product of collaborative activity. Old-fashioned editing played an absolutely critical role. At one point my publishers threw the draft back at me and said it needed rewriting. They were right. Not one of my online collaborators had the confidence to do that; they mainly wanted to encourage me. I spent an awful lot of time working at the text on my own, trying to make sense of all the information. In the final throes of drafting my wife, Geraldine Bedell, a much better writer than I, spent hours improving my hackneyed prose. Every semicolon you see is due to her.

So my own little experiment in writing a book in a more open way shows there is huge potential to engage more people in developing and debating ideas. But that does not mean we can dispense with professional writers and editors to help craft a text. The trick, as the book argues, will be to find the right ways to combine professional and amateur, open and collaborative ways of working with more traditional and closed approaches.

When I was about to put the first draft online I tried to explain what I was doing to Polly Coles, a friend of my sister-in-law. She thought for a moment and replied, 'So the point isn't really to write a book, it's to have a conversation.' That is the point. The book is both the product of conversation and a means to continue it. What matters is the way the book provokes and sustains that conversation. Most of the value resides in that conversation, in how people adapt and appropriate the ideas for themselves.

What I have sought to do with this book is to open up the normally closed process of drafting. Normally a book appears in perfect shape in hardback, then a few months later in paperback. What I have tried to do is to show that a book can start its public life much earlier than that, with various drafts available online: a kind of prequel to the finished work. As a result the book should have a longer, more diverse life once it is published. As well as this physical copy you can download the first three chapters from my website; a wiki version is available where people can add their own comments, additions and references; a four-minute video summarising the argument is available on YouTube along with an animated PowerPoint; on my website you can also find, should you wish to look at it, lots of discarded material that was in earlier drafts but did not make the final cut, including material on collaborative creativity in cities and a profile of shared-approaches innovation on some Norwegian islands. There is also a list of the various titles I considered and then rejected.

Having dipped into this book you can join in that conversation, voice your view, add your bit. You do not have to visit my site to do so. But if you want to make sure I know what you think you can visit www.charlesleadbeater.net to add your comment.

Charles Leadbeater
December 2007

PROLOGUE: THE LEVELLING

The young boy is seated on the edge of his rumpled bed, his slim figure framed by the sunlight pouring through his bedroom window, head bowed, face shaded by his baseball hat, poised over his electric guitar. His fingers are the star of the show, working the fret effortlessly, at high velocity, picking out the complex reworking of Pachelbel's baroque one-hit-wonder the *Canon in D Major*. Originally composed for a harpsichord, bass and three violins, the Canon has become a staple diet of weddings and television commercials. Funtwo, as the diminutive virtuoso guitarist is known on the web, is playing an arrangement for electric guitar created by a prolific Taiwanese amateur guitarist, Jerry Chang. Chang's video of himself playing the Canon on his band's website inspired the largely self taught Korean guitarist Jeong-Hyun Lim, (aka Funtwo), to learn the Canon, using a musical score and backing track downloaded from Chang's site. Lim uploaded his bedroom based effort onto a Korean music site called the Mule, from where it was picked up by 'guitar 90', a regular contributor to YouTube. Guitar 90 wanted more people to see Funtwo's amazing playing. The grainy video lasts 5 minutes and 21 seconds. As of late 2008, almost three years after the video was first posted on YouTube, 51 million people had each watched Funtwo – that's about 272 million minutes of viewing time.[1]

The significance of the 'Funtwo' phenomenon is most clearly revealed by asking what it did *not* take for him to become a worldwide star. Imagine for a minute that Funtwo had taken a different route to find his audience, and gone to an established global media organisation, such as the British Broadcasting Company to air his video.

Funtwo's first challenge would have been to find the right person to talk to among the myriad channels, departments and controllers, the bewildering maze of titles and hierarchies. Then he would have had to hassle, wait, plead and beg to get an appointment with a commissioning editor – at the BBC they are known as Controllers. Getting that appointment would have been tricky because Funtwo did not go to university with any of the Controllers and is not a member of the Groucho Club, the media watering hole in London's Soho. But suspend disbelief for a moment and imagine Funtwo made it through all these hoops and got into see one of the all powerful Controllers. This is perhaps how the conversation would have gone:

Controller: So you want to make a video, what's it about ?
Funtwo: Of myself playing the electric guitar.
Controller: I see – and are you planning to make this video in a studio with sound and lights?
Funtwo: No I thought I could just use my bedroom. It's quiet and sunny.
Controller: And if you are playing who will be shooting the video, where are the director, sound recordist and camera man? Are you working with a television production company we are used to dealing with?
Funtwo: No I thought I could just set up a video camera on a tripod and point it at myself. It's fairly simple.
Controller: Do you have *any* experience directing or performing.

Funtwo: None.
Controller: And how long do you intend this 'video' to be?
Funtwo: Quite short, I only want about 5 minutes, 21 seconds.
Controller: Ah, that's a problem you see because the shortest programmes we do are really about 30 minutes. We'd really need something a bit longer. Do you have an idea when this video would be ready for transmission?
Funtwo: Well if I make it this afternoon I was hoping we could show it later this evening.
Controller: Ha ha, how charming. Let me explain the facts of life to you. You see, our schedules are full for the next nine months. Realistically we could not show anything for another year, at the earliest. And we deal in very large audiences here, which is why we'd really need a celebrity angle, a well known presenter and a format – a bit like *Ready Steady Guitar* or *Strictly Guitar*. Sorry. Final question: I don't want to sound unduly sceptical, because I am sure you are really very talented, Mr Two but how many people do you think will want to watch you playing your guitar?
Funtwo: Oh about 51 million.
Controller: Ha ha, very droll.

The point about Funtwo's video is that he did not have to go through those hoops. He could just play, film, upload and share. He does not have to ask anyone's permission.

That is the big change.

Funtwo did not need the approval of the Controllers to create what he wanted and to make it available to millions of other people, just as Jerry Chang did not have to ask anyone's permission to create his new arrangement of the Canon. Funtwo did not have to ask Chang's permission to take the score and backing track. Guitar 90 did not have to ask Funtwo's permission to take his video from the Mule and upload it to YouTube. None of the thousands of people who have created their own versions of the video, by impersonating, imitating and adapting Funtwo's work asked permission. They just did it. Nothing was planned or scheduled months in advance.

As with most things that get big on the web the video was not the work of a lone genius. Funtwo opened a window on a global micro community of classically trained electric guitar players who avidly play, share and comment on each other's work. These guitarists are classic Pro Ams: they play for the love of it, not for money or fame, but they play to extremely high standards, enthusiastically learning from one another. It is now easier than ever for Pro Ams in many fields to create, publish and share content – whether in the form of music, film, software or text. As a result it is also easier than ever for communities to form around these activities, for people to share, create and learn together. And from that something new and potentially far reaching emerges: people can become organised in new ways, at low cost, without many of the paraphernalia of traditional, hierarchical organisations – head offices, layers of bureaucracy, departments, job titles and so on. That capacity for collective self-expression and self-organisation creates new options for us to become organised, to get things done together in new ways.

Boulders and pebbles

Imagine surveying the media, information and cultural industries in the mid 1980s, industries that provide most of our information and entertainment and so filter our access to the world around us and shape how we make sense of it. The scene would have resembled a large, sandy beach, with crowds organised around a few very large boulders. These boulders were the big media companies.

The boulders came into being because media had high fixed costs – print plants for newspapers and studios for television. They were closely regulated and resources, like broadcast spectrum, were scarce. All that created high barriers to entry. Anyone trying to set up a significant new media business could be seen coming from a long way off. Rolling a new boulder onto the beach took lots of people, money and machinery.

In the mid 1980s an entrepreneur called Eddie Shah tried to roll a boulder onto the British beach by setting up a national newspaper based in northern England. That provoked a protracted national strike. Rupert Murdoch caused controversy by moving his boulder – production of his News Corporation newspapers – from one part of London to another. That caused another lengthy dispute. Channel 4 caused a stir by becoming a new boulder on the beach, one which eventually spawned several others in the form of independent production companies. Then in the 1990s a French company, Vivendi, came along with a plan to merge a lot of the boulders together on both sides of the Atlantic. That didn't work. In the UK several commercial television companies – Granada and LWT – have joined together to create a single very large ITV boulder. The big advertising agencies – WPP and TBWA – are other boulders. Until recently boulders were the only business in town.

Now imagine the scene on this same beach in five years time. A few very big boulders are still showing, but many have been drowned by a rising tide of pebbles. As you stand surveying the beach every minute hundreds of thousands of people come to drop their pebbles. Some of the pebbles they drop are very small: a blog post or a comment on YouTube.

Others are larger such as a video like Funtwo's or a piece of code for a complex open source software programme like Linux. A bewildering array of pebbles in different sizes, shapes and colours are being laid down the whole time, in no particular order, as people feel like it.

Pebbles are the new business. The new kinds of organisations being bred by the web are all in the pebble business. Google and other intelligent search engines offer to locate just the pebble we are looking for. Increasingly Google will be offering to organise more and more of the unruly beach for us. Wikipedia, the free online, user generated encyclopedia is a vast collection of factual pebbles. YouTube is a collection of video pebbles; Flickr is an album of photographic pebbles. Social networking sites such as Facebook, MySpace and Linked In allow us to connect with similar pebbles who are friends or people with shared interests. Twitter, the micro blogging service, allows people to create collections of lots of really tiny little pebbles. Oddly some of the tiniest pebbles seem more powerful than the biggest boulders. In the week that Prime Minister Gordon Brown created his own YouTube channel, a young videoblogger known as charlieissocoollike set up his channel to post his views on the world. A year later the Prime Minister's channel had perhaps 6,000 subscribers while the teenage video blogger had 85,000 subscribers.[2]

There is still a lot of business in serving the boulders that remain, providing them with content, finance, advice and ideas. The boulders still employ a lot of people, but the dynamic growing business is with the pebbles. Of course, the information and media businesses are right at the forefront of the transition from boulders to pebbles because the web so directly affects them. Yet even more traditional sectors will feel the pull of the pebbles in time, not least because the consumers and workforce of the near future will have grown up using the social web to search for and share ideas with one another. They will bring with them the web's culture of lateral, semi-structured free association.

This new organisational landscape is taking shape all around us. Scientific research is becoming ever more a question of organising a vast number of pebbles. Young scientists especially in emerging fields like bioinformatics draw on hundreds of data banks; use electronic lab notebooks to record and then share their results daily, often through blogs and wikis; work in multi-disciplinary teams threaded around the world organised by social networks; they publish their results, including open source versions of the software used in their experiments and their raw data, in open access online journals. Schools and universities are boulders, that are increasingly dealing with students who want to be in the pebble business, drawing information from a variety of sources, sharing with their peers, learning from one another. Most significantly, perhaps, Barack Obama made it to the White House thanks to a campaign which took pebble organisation to new heights. Obama's web based campaign rewrote the rules on how to reach voters, raise money, organise supporters, manage the media and wage political attacks. Mark McKinnon, a senior advisor to President Bush's campaigns told *The New York Times* that 2008 was the year when: 'The campaigns leveraged the Internet in ways never imagined. The year we went to warp speed. The year the paradigm got turned upside down and truly became bottom up rather than top down.' Much of this upheaval took place on YouTube where 'Yes, We can' a video (by will.i.ams) of Obama's words set to music garnered 11.3 million hits, closely followed by 'The Obama girl', a video about a girl with a crush on the candidate which attracted 10.3

million viewers. All around us in myriad ways, many of them very small scale and local, people are pooling their pebbles to find things out, make an impact, and get things done.

And in the midst of all of this companies are trying to figure out how to make money from pebbles. Boulders may be old, heavy and cumbersome, but they have tried and tested ways of making money, even if they are in decline. In contrast, it is much more challenging to work out how to make money from collections of pebbles.

There might be individual pebbles that are particularly beautiful, brightly coloured, finely shaped that are difficult to find. These highly valued pebbles are like popular computer games. Some investors might be interested in pebbles that do a particular job – skimming stones – such as music software like Sibelius and Garage Band, perhaps, they are tools people can use to create content. There might be value in bringing together pebbles in formations, a line of white pebbles, say, or a collection of solely round pebbles – this is the space for social networking companies. And finally then there are people who come to the beach with large containers and gather pebbles together to form very large objects that, from a distance, look like boulders. These containers – YouTube, Flickr, Wikipedia – are ways for people to pool their content, to get it seen. It's easy to mistake these collections of pebbles for boulders. They look very similar but actually they are made up from very different components. Boulders have a very large internal mass in relation to their surface area. They are bigger on the inside than the outside, which is why they tend to be so introspective and internally focused. Pebbles, by contrast, have a very large external surface area compared to their mass. There is relatively little inside a pebble compared a boulder. That is why pebbles tend to be more outward looking.

In writing *We Think* I dropped my own pebble on the beach, with the first draft freely available online, then the eventual book and the four minute YouTube video that goes with it. My video has not reached the dizzy heights of Funtwo or 'Obama girl', but in its first nine months it garnered about 100,000 hits, from people all around the world. For a while it was in the top ten education videos on YouTube in several countries. It got my ideas to a much wider, younger audience than the book.

It also exposed me directly to the vagaries of the chaotic, swarming life among the pebbles. One of the first responses came from astriQn who commented: 'But seriously Charles what colour is the sky on your planet? This utopian cretinism … is heroically fatuous thinking.'

I sat and looked at this comment for a while, at my desk in my office, which is in our garden. One benefit of the civilised, controlled world of the boulders is that people do not go around calling one another utopian cretins, at least not to their face. So I walked across to the house to ask my wife what I should do. She was unsympathetic: 'It's the Internet, what do you expect?'

It did not take long for more comments to arrive, in turns sceptical, supportive, dismissive and grateful. Many shared astriQn's scepticism about the economics of the collaborative web. Shiftfacesam, for example, wondered how my maxim 'you are what you share' would help to put bread on the table. Spaclines chipped in: 'I completely agree with you, philosophically, now I just need to get building societies, utilities and supermarkets to accept my shared ideas as their currency.' Notebookplus put the problem succinctly: 'If we share how do we trade? We need to trade to live.'

Indeed the scepticism extended beyond economics to whether the Internet was a good thing at all. Idd405 commented:

'The bad thing about the Internet is that anyone can spew out false and biased information and people will believe it because it is harder to find valid information.' Within minutes rorypart shot back: 'There is more valid information on the web than in countless libraries.' Crock703 took a different tack: 'The Internet is not a real community it's faceless.'

And many doubted whether human nature would allow us to take the collaborative and creative opportunities offered by the web. DannyLordLoss spoke for several contributors: 'It's a good idea and all, but the human race couldn't pull it off, too much greed.'

Just as many people, probably more, were enthused by the ideas. Medhue captured many of these contributions when he said: 'I feel the Internet is the most freeing thing in the world today. Forcing governments of the world to become more honest. The possibilities for creative people are endless.' Wickedweekend summed it up: 'Who shares the most gains the most.' AdamCromier and HueandCryMusic argued there were business models that would allow people to make money while also sharing ideas and information.

The debate bobbed back and forth, every sceptical response drawing a supportive comment, reflecting both the wider discussion in society and inside our own heads, about whether the Internet is a force for good or ill. The debate over the web's impact on our culture, economy and politics is not a simple argument between good and bad, for and against. It is a complex, still unfolding, mass experiment in which no one is in charge. There are likely to be some very dark corners to this vast laboratory. Plans cannot be dictated in advance in a world where a child in a Korean bedroom can reach 51 million people. During the US Presidential election campaigners said voters were more sceptical, more questioning, better informed, and better able to search for information on the Internet themselves. The flip side, however, was that more voters were prey to gossip-mongering and rumours. The news cycle, once organised around getting stories onto the evening television news bulletins and into the following morning's newspapers, is now more frequent, hyper-accelerated and hyper-connected. There are more opportunities to question issues but less time to think about them. Below are five different but often overlapping views of the future of the Internet. There is nothing wrong in holding more than one of these views at the same time.

Five futures for the internet

One view is that the net is overblown. It is just a tool to do what we've always done but it allows us to do it more quickly and to reach a larger audience. eBay is just a flea market and an auction system – nothing new there – but with the scale and speed of the Internet applied to them. Amazon is just a different way get goods delivered to your door: big deal. Bill Gates once took this position, dismissing the Internet as a side show. In the backlash against the Internet which followed the dotcom boom of the late 1990s many adopted a similar stance. At the time, some academics suggested the web would be a bit like CB radio: a back channel for rough and ready amateur communication.

This view is partly true, of course. The net is a tool that allows us to do things we already do but in a different way – my kids watch television programmes but on YouTube. Yet the web is more than just a tool: it allows people to do so much more than simply watch, buy, sell, click. Microsoft has since changed its stance. In 2008 it announced a new web-based strategy that would allow 'cloud' computing with computer users accessing many shared programmes running externally in a cloud on the web – email, calendars, shared documents – running externally rather than installing the software on their own computers. Even Microsoft realises the future value in connecting up the pebbles.

A second view is that the Internet might well have a big impact on society but it will take a lot longer to work through than the super-optimists argue. Technical change usually does. The big productivity gains from technological innovation come when the technology itself becomes so dull that it is integrated seamlessly into daily life. Think of the way we effortlessly flick on a kettle or turn the key in a car's ignition. Successful technologies are used without a second thought. The web has become more accessible, reliable and user friendly but it is still some way off from being second nature for most people. A prime exponent of this sceptical view is David Edgerton, a historian of technology. In his book *The Shock of the Old* Edgerton argues that technical change is rarely revolutionary: older technologies – waterwheels, horse-drawn carriages, radios – take a lot longer to fade away than we imagine.

This view might offer some comfort to people who are worried about the web: we might have more time to get used to the web than we initially thought. The changes it brings might be evolutionary rather than disruptive. Yet if the Edgerton view is correct it would also mean that the changes we have seen in just the first decade of the mass adoption of the web – the complete upheaval in the music recording industry, the savage decline in US newspapers, the disappearance of many youth magazines, the quick creation of new media giants like Google – these might just be the tip of an iceberg. We have another fifty years of change of this kind to come and the scale of the upheavals may be even greater as the technology becomes widely adopted and gains momentum.

A third small but vociferous group are people who say the web is already having a big impact on society and it is mainly bad for us . The chief proponents of this view are the polemicist Andrew Keen in his book *The Cult of the Amateur*, Nicholas Carr in his thoughtful *The Big Switch*, Larry Sanger, one of the co-creators of Wikipedia and the brain scientist Susan Greenfield.

These critics worry the web is uprooting the authority of experts, professionals and institutions which help us to sort truth from falsehood, knowledge from supposition, fact from gossip. Instead the web is licensing a cacophonous mass in which it is increasingly difficult to discern the truth as experts themselves are drowned out by low grade amateurs. This scepticism is echoed by many professions who feel threatened by the web: journalists, teachers, academics and librarians among them. The boulders might have been cumbersome but they filtered the good from the bad before it was published. In web world things get published first and then filtered afterwards, depending on people's reactions to them.

A related worry, articulated by Nicholas Carr and Susan Greenfield, is that our dependence on the web and computers is eroding our ability for independent thought. 'Google is making us stupid,' is Carr's catchphrase. We accept the answers the search engines supply without really analysing what they mean or where they came from. Screens make our thinking dependent on stimulation, according to Greenfield. We are no longer self-starting, critical thinkers. Greenfield also worries that the web is relentlessly eroding our sense of privacy as young people live out their lives in public, online, unable to form and protect a stable sense of identity.

My view is that much of this is alarmism, arising from a rosily nostalgic view of what the old world was like. It also under estimates people's capacity to self organise, to sort fact from fiction, the useful from the fraudulent. The web provides more opportunities for participation, critical thinking and search than say sitting in front of a television or simply copying down facts from a blackboard. Far from losing a sense of identity younger generations growing up with the web seem both more individualistic and more collaborative than their elders. That is not to say that these critics do not raise important points, but they are qualifiers, not the main story.

The fourth group argue the net will be mainly good for us. The members of this group, however, differ over why and how the net will be useful for society.

The libertarian, free market wing believe the Internet is creating more diversity and choice, resulting in faster, frictionless markets and an abundance of free culture. In fact, the web is no less than a capitalist cornucopia. Chris Anderson, the editor of *Wired* and author of *The Long Tail* is the cheerleader for this camp.

The communitarian optimists take a contrary view. They see in the Internet the possibility of community and collaboration, commons-based, peer-to-peer production, which will establish non-market and non-hierarchical organisations. It is not opening a new stage of capitalism and the market but laying the seeds for alternatives to both. *We Think* stands in this camp along with Clay Shirky's *Here Comes Everyone*. The most comprehensive statement of this collaborative point of view is Yochai Benkler's *The Wealth of Networks*.

The real measure of the Internet's impact and value is

whether it provides radically different options for how we organise ourselves: generating knowledge, sharing ideas, creating culture, making decisions together.

People want meaningful opportunities to participate and contribute, to add their piece of information, view or opinion. They want viable ways to share, to think and work laterally with their peers. They are searching for collaborative ways to get things done. When these three come together – participate, share, collaborate – they create new ways to organise ourselves that are more transparent, cheaper and less top down: structured, free association.

Not even all web optimists share these views. Some reject both high-minded communitarianism and rampant free market liberalism. Best represented by David Weinberger, author of *Small Pieces Loosely Joined*, this group argues the web cannot be easily parcelled into an ideological position because it is about *everything* that makes up everyday life. The web comes with all human motivations in play – greed, vanity, kindness, cruelty. Only a part of it will ever resemble a worthy collective dialogue. That may be true. My view, however, is that for the web to be really significant we have to do more than create lots of pebbles. We also have to find ways to join them up effectively and imaginatively, that is when the web becomes really powerful.

Finally a small group argue the net has been largely good so far and has huge potential but it may turn bad. As the Internet grows it produces more of its own pollution – spam, malware, surveillance, invasions of privacy, trivia. The danger is that if there is too much chaos and abuse then the net will get clogged up and eventually people will turn back to corporations or governments to sort it out. In this sense the Internet is its own worst enemy. As it stands, the web is a vast space governed by extraordinarily loose, ramshackle forms of self-governance. If we mess this up then the ensuing chaos will feed an appetite for more traditional and reliable forms of control. The boulders will crush the pebbles.

The task is to save the Internet from its own self-destructive tendencies. The chief exponent of this view is Jonathan Zittrain in his book *The Future of the Internet and How to Stop It*. Zittrain's point is that we are just at the beginnings of the web and this early experimental period may be no more than a passing moment. Even in its short life the Internet has already had several incarnations such as the Information Superhighway (which no one refers to anymore). Amid the buzz of Web 2.0 few people talk about e-commerce. Many are already starting to talk about the 'cloud' as a new paradigm. The Internet is ever changing and developing. The Internet that was created by a peculiar mixture of academics, hippies and geeks in the 1960s may not last. The Internet as we know it was built on an ethic of mutual self-help with the all purpose, reprogrammable personal computer as the main point of access. Yet if the personal computer is replaced with less open, adaptable devices – like the Apple iPhone – then it will be far harder for hackers, amateurs and kids to innovate at the edges of the system. The Apple personal computer helps people to create; it feeds the mutual, self-help, hacker ethic of the web. The Apple iPhone, the iPod and iTunes are very different. What you can do depends on what Apple allows you to do. The Apple iPhone is a seductively dangerous little tool. If the future web is accessed through constrained mobile devices then it will be considerably less free and easy than it has been to date.

One of the most striking things about the Internet is that

we call it *the* Internet – a singular shared space to which we can all go. In the early days, the likes of Microsoft, Yahoo, Compuserve and AOL failed in their attempts to carve the web up into separate domains, so called 'walled gardens'. Yet there is no reason why versions of these walled gardens might not be recreated in future. Apple is busy applying its reputation for outstanding design to create one around iTunes. The net of the future might dissolve into many such managed spaces. Google's ambition to organise the world's information means us relying on it to organise the 'cloud'.

Zittrain warns us that the collaborative, open web we inherited from the academics and geeks might be no more than a passing phase. Like previous experiments in communal living and self-organisation it may well not last. Indeed the best guide to what might be in store for the web could be found in a few revolutionary years of English history more than 350 years ago.

The new Levellers

On 1 April, 1649 in the midst of a revolution which had seen the King executed, a new democratic army created and the social and political order shaken to the core, Gerrard Winstanley led a group of masterless, landless, virtually starving men to the top of St George's Hill just outside London and started digging, to cultivate unused land and to feed themselves. The Diggers as they became known were one of the most radical sects of the English Revolution, putting into practice Winstanley's idea for a self-governing, cooperative, productive community.

Winstanley, the chief intellectual inspiration for a larger movement known as the Levellers, set out to challenge the social order of the day by levelling relations of power, knowledge and economics. Winstanley's economic argument was that mutual ownership of under-used land would boost productivity, raise incomes for the majority, feed the hungry, provide work for the unemployed and share wealth more equitably. The land, Winstanley argued, was a common treasury that should be held in common ownership. He did not seek to overthrow private ownership but just to limit its reach. He wanted unused land to be cultivated communally. Winstanley wanted political power levelled, with democracy extended, so that all people, not just men of property, could have a say in electing officials. Underlying this was a challenge to the authority of the clergy as the font of all truth and knowledge. Far ahead of his time Winstanley argued for mass education of boys and girls up to the age of eighteen. He wanted universities to be freed from the control of the church and for the creation of a system in which localities shared with one another new ideas to harness innovation. To spread the word the Levellers created one of the first ever mass circulation newspapers called *The Moderate*. Winstanley believed a modern society, governed by the law of freedom rather than the yolk of monarchy and aristocracy, should be based on as many people as possible having access to and laying claim to knowledge.

In the words of a famous Leveller song, *The World Was Turned Upside Down*, for a few months the bottom of society became the top and the poor saw an opportunity to take power into their own hands. As Christopher Hill put it in his history of the period, Winstanley's ideal was 'a society of all-round, non-specialists helping each other to arrive at truth through the community.'[3] Sounds a bit like the Internet to me.

Winstanley set out guidelines for collective cooperative working in a pamphlet know as *The Digger's Covenant*. It was intended to be an alternative system of production to the emerging market and old style feudalism: Diggers promised to work together and eat together, rather than work for another (a lord) or to live on wages (and work for an employer.) Only by freely labouring with one another would there be food for all. Common property, shared production and religious toleration were the conditions for liberty and piety, to live a free but moral life, according to Winstanley.

The groundswell of hope being invested in the web's capacity to create more collaborative, less hierarchical communities, in which knowledge and power are more evenly distributed, comes from a line of radical, utopian thought which started with Winstanley and the Levellers. For more than a decade *Wired* magazine and many others have being telling us the world is being turned upside down by a technology that empowers the edge of society at the expense of the centre, the bottom at the expense of the top. The Web 2.0 optimists are the new Levellers. They argue we are in the midst of a new great levelling thanks to the revolutionary, decentralising technologies of the web. The technology might be new, but most of their ideas are not. They were first foreshadowed by an English revolutionary digging the land in 1649.

And therein lies the challenge for today's Levellers. Most of the original Levellers' ideas to extend democracy, education, welfare, communication, freedom of religious observance, gender equality – eventually came to pass. The Levellers had a profound influence on the American revolution. A former Leveller gave the sermon bidding farewell to the Pilgrim Fathers from Plymouth. Leveller songs were played to celebrate the British defeat in the War of Independence. Yet it took several centuries for the Levellers' ideas to become widely accepted. In their own time they were a failure, their utopian communities quickly collapsed and disbanded because they could not support themselves economically and provide a viable alternative to the established order. Most basically, they could not put bread on the table. They were brave, inspiring but ultimately doomed experiments. Within months of the Diggers establishing camp on St George's Hill the Monarchy had been restored to power, the army turned against the people and landlords ruled the land once again. Diggers and Levellers were killed, imprisoned or driven into hiding.

The next few years will determine whether the latter-day new Levellers of Web 2.0 suffer a similar fate, their inspiring collaborative schemes brought down by their own frailties, providing the opportunity for a sweeping restoration of traditional top down power in the form of corporate control and government regulation. Winstanley, one of the most radical voices in English history ended his life embracing everything he once despised, dying as a landowning Church Warden and Chief Constable.

Whether we miss this opportunity to create more equitable, collaborative and participative ways to organise ourselves will be one of the big stories of the next decade. *We Think* is an attempt to chart how the new Levellers might avoid the crushing defeat inflicted upon the original Levellers. Vital to that will be the scale of our ambition to realise the shared potential the web offers.

Web creator, Tim Berners-Lee, an English radical following in Winstanley's footsteps says: 'The danger is not that we ask too much of the Internet but too little, that we turn it into just another piece of kit when it could be so much more significant than that, a new platform for how we could organise ourselves, to find knowledge together, to work out what is true and to decide together what we should do about it.'

We are living through a great levelling brought on by the power of the web. What we make of it is still, thankfully, up to us.

Charles Leadbeater
December 2008

1
YOU ARE WHAT YOU SHARE

If you are not perplexed, you should be. As the web becomes ever more widespread, infiltrating our lives and shaping what we think is possible, we are increasingly unnerved about what we might have unleashed. Will the web promote democratic collaboration and creativity? Or will it be a malign influence, rendering us collectively stupid by our reliance on what Google and Wikipedia tell us being true, or, worse, promoting bigotry, thoughtlessness, criminality and terror? How will it change the way we think and behave and what will its growing domination of the world of information and ideas do to us? Clearly there is enormous potential.

Thanks to the web, more people than ever can exercise their right to free speech, reviving democracy where it is tired and inspiring its emergence in authoritarian societies, from Burma to Vietnam and China. In theory, the web should be good for democracy. Yet often this extension of free speech seems to produce little more than a babble of raucous argument that rarely turns into the structured and considered debate essential for democracy to thrive. Bloggers cannot overthrow authoritarian rulers on their own.

As a consequence of the web, our freedoms have exploded – not just to shop for cheap, last-minute deals, but to be creative with tools that help us to express ourselves through writing, making videos, composing music. More people than ever now have such tools. On YouTube, for example, you can see videos made by performance artists which have attracted millions of viewers. Ideally the web should spread the freedom to express ourselves creatively. Yet the web also expands the scope for surveillance, not just by the state and corporations, but also by our peers and friends. Every move we make on the web leaves a little wake that can be tracked. Any indiscretion of youth could come back to haunt us, thanks to a user-generated surveillance system of social networking in which everyone is keeping an eye on everyone else. Young British tennis stars were stripped of their status after ill-advised revelations on their Facebook sites. Nothing seems to be private any more, and that surely cannot be good for freedom.

The web also promises to be good for equality. Barriers to information and knowledge are falling fast. Information and knowledge are vital inputs into everything that matters, from education to creating new drugs or devising clean energy systems. Thanks to the web, more people than ever should have access to knowledge, and that should help education and innovation among the poorest people in the world, those who can least afford schools, libraries, universities, laboratories. The web, in theory, should be good for equality.

Yet the web most rewards those who are already well connected, by allowing them to network together, reinforcing their privilege. Economically the web seems to destroy as much as it creates, and many wonder whether on balance this leaves us better off. As more of us turn to the web for news, information, entertainment and conversation, for example, we turn away from newspapers, television, film, libraries, bookshops. That may liberate us from the control of a cultural élite – editors and publishers, critics and commentators who used to oversee what we read and thought. Yet the orgy of user-created content the web has attracted might also rob us of high-quality journalism and literature, film and music, as the institutions that train and employ professionals find their economic foundations eaten away. In the US, the spread of social networking sites such as Craigslist is destroying the market for local newspapers; who is to say what the long-term impact of that will be on communities that will no longer have such a focal point? We may come to rue the YouTube cultural revolution if it banishes the gatekeepers of quality and culture to the digital wastelands. No amount of amateur blogging will make up for well-trained and -funded investigative journalism that makes politicians quake, probing the depths of scandals the powerful want to keep quiet.

Many people are deeply uncertain about whether the world the web is creating will leave us feeling more in control of our lives or less. On the one hand the web is the source of our most ambitious hopes for spreading democracy, knowledge and creativity. It ought in principle to give us untold capacity for solving shared problems by allowing us to combine the knowledge and insights of millions of people, creating a collective intelligence on a scale never before possible. But on the other hand the web is also the source of some of our most lurid fears: it has already become a tool for stalkers, paedophiles, terrorists and criminals to organise networks for shadowy purposes beyond our control.

The web's extreme openness, its capacity to allow anyone to connect to virtually anyone else, generates untold possibilities for collaboration. It also leaves us vulnerable to worms, viruses and a mass of petty intrusions. The more connected we are, the richer we should be, because we should be able to connect with people far and wide, to combine their ideas, talents and resources in ways that should expand everyone's prosperity.[1] But the more connected we are, the easier it is for small groups to cause enormous disruptions, by spreading viruses, real or virtual. The web enables small, dispersed groups to collaborate in ways that were previously impossible. That might be great for the small community that trades car parts for old Citröens, or for those who want to play poker against one another. It could be dreadful if it empowers a small group of fanatics to explode a dirty bomb in a major city. The more connected we are the more opportunities for collaboration there should be, but also the more vulnerable we become.

The web's critics argue that it will corrode much of what is valuable in our culture, which rests on learning and expertise, professionalism and specialism. All too easily, social networking could license an obtuse group-think. It will be harder for dissenters to diverge from the party line of their peers. That is likely to amplify errors and prejudices rather than correct them, to aggravate bias and sustain falsehoods that should be challenged. As the Internet encourages more people to disappear down their cultural boltholes, seeking out others who share their views, what little is left of our common culture could fracture as people pursue their own, separate conversations. Music and film companies complain that the web is destroying established business models vital to allow investment in talent. The optimists describe the web as a conversation. Yet much of the web seems raucous and unruly, more like a bar-room brawl than a moderated discussion.

Every interaction we have with the web is laced with uncertainty. How can we be sure what is true when a free-form encyclopaedia compiled by anonymous volunteers – Wikipedia – gets more traffic than the BBC? What is to be counted as real in a world where some people spend the equivalent of a day a week in virtual worlds such as Second Life and World of Warcraft being avatars or alts? Or take the apparently simple question of what it means to be someone's 'friend'. In the world before social networking became the new religion, 'friend' was a term reserved for one of a small band of people to whom you were close and on whom you could depend in a crisis. In the social-networking sphere the idea of 'friends' encompasses passing acquaintances, fans and even people you do not actually know. How can the web be good if it so aggressively degrades an idea so vital as that of friendship?

We are reaching a critical phase in the web's development, when we will see more clearly how it will influence society, not just in the rich developed world where it began, but even more importantly in fast-developing economies in Asia and South America, where in the next decade close on a billion people will access it through cheap mobile phones and laptops. What started a few decades ago as an intriguing file-sharing experiment among academics is reshaping culture around the world, changing how we will think and relate to one another. We will look back on the coming decade as a period of unparalleled social creativity when we sought to devise new ways of working together to be more democratic, creative and innovative, potentially on a vast scale. The web could amplify our combined intelligence if only we can find ways to use it to work creatively together. If not, it could lead to anarchy, an anything-goes culture increasingly beyond central control, in which potentially lethal ideas and technologies flow out of the institutions where they were once under the mastery of professionals and into the hands of people who cannot be trusted to use them wisely. We may rue the day we let the genie out of the bottle.

This book is about how we can make the most of the web's

potential to spread democracy, promote freedom, alleviate inequality and allow us to be creative together, *en masse*. The web's potential for good stems from the open, collaborative and even communal culture it inherited from its birthplace in academia and from the counter-culture of the 1960s, combined with pre-industrial ingredients it has resurrected, folk culture and the commons as a shared basis for productive endeavour. The web allows for a massive expansion in individual participation in culture and the economy. More people than ever will be able to take part, adding their voice, their piece of information, their idea to the mix.

Greater individual participation will not, on its own, add up to much unless it is matched by a capacity to share and then combine our ideas. In the last 30 years the spread of the market, the collapse of Communism and the travails of the public sector have elevated private ownership as the best way to organise virtually everything. The spread of the web invites us to look at the future from a different vantage point, to see that what we share is at least as important as what we own; what we hold in common is as important as what we keep for ourselves; what we choose to give away may matter more than what we charge for. In the economy of things you are identified by what you own – your land, house, car. In the economy of ideas that the web is creating, you are what you share – who you are linked to, who you network with and which ideas, pictures, videos, links or comments you share. The biggest change the web will bring about is in allowing us to share with one another in new ways and particularly to share ideas. That matters because the more ideas are shared the more they breed, mutate and multiply, and that process is ultimately the source of our creativity, innovation and well-being. This book is a defence of sharing, particularly the sharing of ideas.

The web matters because it allows more people to share ideas with more people in more ways.

The web's underlying culture of sharing, decentralisation and democracy makes it an ideal platform for groups to self-organise, combining their ideas and know-how, to create together games, encyclopaedias, software, social networks, video-sharing sites or entire parallel universes. That culture of sharing also makes the web difficult for governments to control and hard for corporations to make money from.

In reality, creativity has always been a highly collaborative, cumulative and social activity in which people with different skills, points of view and insight share and develop ideas together. At root most creativity is collaborative; it is not usually the product of a lone individual's flash of insight. The web gives us a new way to organise and expand this collaborative activity.

The factory made possible mass production, mass consumption and with that the industrial working class. The web could make innovation and creativity a mass activity that engages millions of people. The developed world in the 20th century was preoccupied with organising and reorganising the mass-production system, its factories, industrial relations systems, working practices and supply chains. Our preoccupation in the century to come will be with creating and sustaining a mass innovation economy in which the central issues will be how more people can collaborate more effectively in creating new ideas.

As the web shapes and colours many more aspects of our lives, it will provide us with a new way of thinking, a set of reflexes for how we should organise ourselves. For the generations growing up with social-networking sites, multiplayer computer games, free software and virtual worlds, the reflexes learned on the web will shape the rest of their lives: they will look for information themselves and expect and welcome opportunities to participate, collaborate, share and work with their peers. The web will slowly reframe how we see the more material aspects of our lives fitting together. The factory encouraged us to see everything through the prism of the orderly production line delivering products to waiting consumers. The web will encourage us to see everyone as a potential participant in the creation of collaborative solutions through largely self-organising networks. But that will only come to pass if we can organise our shared intelligence ourselves. How we do that is the challenge this book addresses. A couple of examples of what could be possible might help to explain.

In late July 2004, in the closing frames of cinema advertisements for Halo 2, the science fiction computer game, a website address – www.ilovebees.com – flickered across the screen. Over the following few days, thousands of Halo fans and others intrigued by the address visited the site, which appeared to belong to an amateur beekeeper called Margaret, who had disappeared. Her honey-based recipes had been replaced by 210 global positioning system co-ordinates. Attached to each set of co-ordinates was a time of day, these times spaced out at four-minute intervals over 12 hours. A message warned that 'the System' was 'in peril' and a clock was counting down to a date which proved to be 24 August. At the bottom of Margaret's homepage was the question 'What happened to this page?' and a link to a blog written by Margaret's niece Dana, who exchanged about a hundred emails with visitors before herself disappearing without explanation.[2]

collaboration
DIVERSITY
independence
CREATIVITY

That was it: no instructions, no rules, just a puzzle to solve, a seemingly complex set of numbers and a ticking clock. Over the next four months, 600,000 people – mainly US college and high-school students – set out to solve the mystery of Margaret's web page by finding out what the co-ordinates meant. What unfolded was a striking display of mass collaborative creativity and intelligence. The participants in I Love Bees started to throw around ideas and share information about what the co-ordinates meant. They set up blogs and bulletin boards, websites and instant message groups. But they did not simply gather, publish and share information: beneath the blizzard of emails and blogs there was a discernible order. They started to sift, sort and analyse the information together. They debated theories about what the co-ordinates stood for, formed plans, and split into teams to pursue different avenues of inquiry. Eventually, after many failed attempts to work out what the co-ordinates meant, they created a theory that all the players shared, and in the final stages they decided, *en masse*, how thousands of people should take co-ordinated action. They achieved this without knowing one another and without anyone's being in charge. There were no bonuses on offer, nor any of the other incentives we assume are needed to get people to work. The participants were highly organised without having much by way of an organisation.

The I Love Bees game, designed by Californian company 42 Entertainment, had its roots in flash mobbing, a form of public performance art which had started in New York and San Francisco in 2003. In flash mobs, anything from a handful of people to several thousand, who have organised themselves by word of mouth, over mobile phones and via the Internet, gather in a public place, such as at a railway station or on a street crossing, to undertake an apparently bizarre activity.[3]

Jane McGonigal, one of 42 Entertainment's lead designers and a pioneer of flash mobbing, designed I Love Bees to see whether a mob could become a creative force.

In the four weeks after the advertisements were shown, the game designers fed clues to the players through hundreds of websites, blogs, thousands of emails and more than 40,000 MP3 transmissions. These clues were released to players all over the globe, so a player anywhere could find themselves with an important role. The players had to share their evidence to make sense of it. One new clue on Dana's blog, for example, attracted 2,041 comments in just a few days. A popular message board clocked 50 posts every 30 seconds in the first few weeks. In the first 10 weeks of the game, players made more than a million message-board postings. One group of about 4,000 players, known as the Beekeepers, became the core of the community, producing scores of hypotheses about what the co-ordinates might mean. It was the Beekeepers who discovered that at each of the 210 locations spread around the world there was a pay phone.

The game began to come to a head on 24 August, as thousands of players turned up at the pay phones armed with every conceivable piece of digital communication equipment, including databases of players' mobile-phone numbers, camcorders, GPS systems, scanners and satellite phones. As the day unfolded, at the time specified on the list of co-ordinates, the pay phone in question would ring and the player answering was asked a question. If they got the answer correct, which all did, they were played a snippet from a drama about Margaret. The group's task was to put the snippets in the right order by the end of the day and to post the completed work on the web. They succeeded.

That was the first of several tasks set by the puppet masters.

Over the next 12 weeks, the number of co-ordinates and pay phones rose from 210 to 1,000, all around the world. The game reached its climax one Tuesday in late autumn. Shortly after sunrise, the puppet masters started calling pay phones on the US east coast. Whoever answered had to provide a piece of intimate information five words long. The puppet masters then revealed they would call another of the 1,000 pay phones and expect to be told the same five words. The players had an hour to get the five words to everyone else playing the game, all across the world, at all of the 1,000 phones. The puppet masters staged a dozen of these information relay races. In the last of these races the players had 15 seconds to get the five words from the person who answered the first call to the person taking the second call. They never once failed.

The 600,000 players of I Love Bees showed that a mass of independent people, with different information, skills and outlooks, working together in the right way, can discover, analyse, co-ordinate, create and innovate together at scale without much by way of a traditional organisation. Their collaboration was not an anarchic free-for-all; it was organised, but without a division of labour imposed from on high. So if some ingenious west coast games designers can create the conditions in which thousands of people around the world collaborate to solve a trivial puzzle, could we do something similar to defeat bird flu, tackle global warming, keep communities safe, provide support for disaster victims, lend and borrow money, conduct political and policy debates, teach and learn, design and even make physical products?

Whether this hope turns out to be reasonable or hopelessly idealistic may depend on the eventual fate of a global experiment in sharing that is still in progress: Wikipedia. This free, volunteer-created encyclopaedia is revered and denounced in equal measure: worshipped with fervour by its admirers as a wonder of collaborative creativity and pilloried by critics as a licence for anarchy, a platform for half-truths and a free ticket for ill-informed amateurs to gain credence they do not deserve at the expense of knowledgeable professionals.

Wikipedia is the offspring of an ultimately ill-fated collaboration. In 2000, Jimmy Wales, a former options trader, employed Larry Sanger to create a free online encyclopaedia, Nupedia, which would allow anyone to submit an article to be reviewed by expert editors before being published.[4] The seven-stage editorial review Sanger designed proved cumbersome and, as a result, Nupedia grew slowly. The first article – on atonality – was published in the summer of 2000, and Nupedia peaked in the winter of 2001 with 25 published articles. Over dinner on 2 January 2001, Ben Kravitz, a software-programmer, introduced Sanger to the 'wiki', a web page that could be directly edited by anyone with access to it.[5]

Sanger saw how a wiki could help build an open encyclopaedia by allowing writers and editors to work on a shared document. In a memoir of the project's early days Sanger identified the benefits:

> Wiki software does encourage, but does not strictly require, extreme openness and decentralisation: openness since page changes are logged and publicly viewable and pages may be further changed by anyone; and decentralisation, because for work to be done, there is no need for a person or body to assign work, but rather, work can progress as and when people want to do it. Wiki software also discourages the exercise of authority, since work proceeds at will on any page and on any large, active wiki, it would be too much work for any single overseer or limited group of overseers to keep up.

Sanger wanted to revitalise Nupedia, but Wales saw a more radical possibility: to create an entirely open, highly collaborative approach to knowledge. Wikipedia's domain name was purchased on 15 January 2001. By the end of that month there were already 31 articles; by March, 1,300; and by May, 3,900. Sanger left the project as an employee in 2002 and has since become one of Wikipedia's sternest critics. In 2007 he launched Citizendium, a competitor online encyclopaedia, which aims to bring together experts and amateurs.

Wikipedia's advocates believe wiki culture encourages shared creativity and responsible self-governance. Critics say it licenses an anything-goes approach to knowledge. Students, they allege, assume everything on Wikipedia is true. Rather than think, question and explore for answers themselves, they cut and paste the answer they get from Wikipedia. The critics argue this licenses intellectual laziness on a grand scale as we devolve to Wikipedia the responsibility for telling us what is true and what false. A few people involved in Wikipedia might think for themselves more; the overall result is that the general public think for themselves less.

Adjudicating these claims is tricky because Wikipedia is still developing. What is beyond doubt is that it has sustained remarkable growth. From 31 articles in English in January 2001, Wikipedia had a year later amassed 17,307, rising to almost a million by January 2006, and 1.5 million by 2007, when the number of articles in all languages topped 6 million. The rate of growth in articles in English between 2001 and 2007 was 5 million per cent, and for articles in all languages 19 million per cent. By mid-2007, Wikipedia had more than 450,000 articles in German and more than 1,000 articles in over 100 languages. Wales says his aim is to create the Red Cross of information: to put the knowledge contained in a large encyclopaedia in the hands of everyone on the planet, for free. As of March 2007, Wikipedia was used by 5.87 per cent of Internet users, compared with 0.03 per cent for the Encyclopaedia Britannica, 1.73 per cent for the BBC news website, 1.36 per cent for CNN and 0.62 per cent for the New York Times. Wikipedia was ranked as the 11th-most-visited website in the world, while Encyclopaedia Britannica languished at 4,449th.

For a long time, Wikipedia had one employee. By 2007 it had five. Wales has invested perhaps $500,000 in the project. Public donations to the Wikimedia Foundation, which runs the site, have become much more significant: in 2006 they totalled $1.5 million. Still, these are very low costs for the creation of something on this scale. Most of the articles have come from people who want to contribute to a shared resource. Their contributions are edited not by experts but by open debate among peers. Behind each entry in Wikipedia lies an extensive talk page which documents all the debate between participants over what to include, change or exclude. The average article has been subject to about 11 edits. By January 2006, about 154,885 people had made more than 10 edits each, 78,308 of these in English.

Yet Wikipedia works only because this mass of contributors organises itself in a very particular way. Most of the editing is done by a relatively small group. In January 2006, for example, while 47,297 people each contributed more than five times to all language editions of Wikipedia, only 7,460 made more than 100 edits each. This sliding scale of contribution is crucial to the project's success, which has come to depend heavily on a core of highly active participants who each look after a set of pages, eliminating vandalism and deciding on corrections. This core group, which resembles the Beekeepers in I Love

Bees, works on the many millions of Wikipedia contributions made by tens of thousands of people.

One early lesson from both I Love Bees and Wikipedia is that creative communities are not egalitarian. Wales describes the community's self governance this way:

> In part Wikipedia is anarchy. Really, no one is in control of the content, its up to people to sort it out for themselves. That also means it is a meritocracy: the best ideas should win out. In part, it is democracy because some things do get voted on. There is also an element of aristocracy: people who have been involved in the community longer, who have acquired a reputation have a higher standing in the community. And then there is monarchy – that's me – but I try to get involved as little as possible.

The most contentious question about Wikipedia is the one that really matters: how good an encyclopaedia is it? Sanger argues that its quality is questionable because its experts do not vet amateur contributions. In an influential online essay cultural critic Jaron Lanier branded it a form of digital Maoism on the grounds that it promotes an anonymous collective account of knowledge that on any subject favours the often inaccurate lowest common denominator. Others allege that Wikipedia licenses gossip and falsehoods to masquerade as truth, because contributions are often not checked fully. The answer is that we do not yet know how good Wikipedia is or will become. Much will depend on how the community organises itself and that may well evolve, giving a larger role to the core to ensure quality and limit vandalism.

Wikipedia is unquestionably more populist in its coverage than *The Encyclopaedia Britannica*. If you look up Barbie in *The*

Encyclopaedia Britannica you will find an article on the Nazi war criminal whose first name was Klaus. On Wikipedia you will find a lengthy, thoughtful and entertaining account of the children's doll. Wikipedia is often good at explaining current and unfolding events: senior BBC executives acknowledge that Wikipedia's account of the 7 July 2005 terrorist bombings in London was as good as the corporation's. And Wikipedia operates on a vast scale: the *Britannica* has 44 million words of content, Wikipedia 250 million.

It would be foolish not to acknowledge that Wikipedia is not perfect. Like all publishers it can make mistakes.[6] On the other hand it is difficult to establish just how serious these mistakes are. A survey by *Nature* magazine asked expert reviewers to compare 42 articles on Wikipedia with corresponding entries in the *Britannica*. Eight serious errors were detected, four from each encyclopaedia. Reviewers found 162 factual errors, omissions or misleading statements on Wikipedia and 123 in the *Britannica*. *Nature* concluded that its survey showed Wikipedia came close to the *Britannica* in terms of accuracy. *Britannica* retorted that it was 30 per cent more accurate, a not insignificant difference.

Yet if Wikipedia is prone to more errors, it also seems to heal itself remarkably quickly and openly. Robert McHenry, the *Britannica*'s former editor-in-chief, derided Wikipedia as a faith-based encyclopaedia by pointing to flaws in an article on Alexander Hamilton, one of the founding fathers of the US constitution.[7] Hamilton's biographers cannot agree on whether he was born in 1755 or in 1757.[8] Wikipedia seemed to have ignored this controversy and plumped for 1755. (Although McHenry did not note this, commercial online encyclopaedias produced by professionals also failed to reflect the controversy.) Within a week of McHenry's attack, however, Wikipedia's self-healing mechanism had produced a reasonably clean version of Hamilton's biography. One academic study found that almost all acts of vandalism in May 2003 were repaired within minutes.[9] As Wikipedia has grown, so more articles – for example those on President George W. Bush, Israel and the Iraq war – have been subject to such repeated abuse and vicious dispute that they have been withheld from public editing. Yet although abuse, self-promotion and vandalism are a growing problem – what would one expect with something that is entirely open and has 6 million articles? – these occur in less than 1 per cent of the total. Invariably, Wikipedia is a good place to start researching a topic, but rarely provides the final word. Its weaknesses would pose a threat to the way we establish what we know only if it were to become a monopoly supplier of knowledge, displacing other sources. That seems extremely unlikely.

The most important point about Wikipedia, however, one that is often overlooked by its parochial, US-centric critics, is this: most people in the world cannot afford to compare Wikipedia with the *Britannica*. They will not be able to afford an encyclopaedia in any form for many years to come. Wikipedia is creating a global, public platform of useful knowledge that will be freely available to any school, college or family in the world, in their own language. In Africa, even where communities do not have access to the Internet, teachers are using copies of Wikipedia downloaded onto CDs. Wikipedia may get the odd thing wrong, but that misses the bigger picture. Jimmy Wales and his community have created a new way for us to share knowledge and ideas at scale, *en masse*, across the world. Wikipedia's message is: the more we share, the richer we are.

As Wikipedia spreads around the world not only does it

carry knowledge, it teaches habits of participation, responsibility and sharing. Wikipedia is based not on a naïve faith in collectivism but on the collaborative exercise of individual responsibility. Wikipedia is one of the most amazing cultural creations of modern times: a global resource of 6 million, volunteer-created articles amassed over six years, with virtually no staff and little funding. Wikipedia is like a vast bird's nest of knowledge, each piece of information carefully resting on another. Yet this is a bird's nest with no bird in charge of where to put each piece. It has almost constructed itself.

I Love Bees and Wikipedia are both examples of We-Think – my term to comprehend how we think, play, work and create, together, *en masse*, thanks to the web. In most fields – science, culture, business, academia – creativity emerges when people with different vantage points, skills and know-how combine their ideas to produce something new. The web provides a platform for us to be creative together on a scale previously unimaginable. It is changing how we share ideas and so how we think.

The phrase *cogito ergo sum*, 'I think, therefore I am', was inscribed onto our culture in 1637 by the French soldier-cum-philosopher René Descartes, announcing a dramatic inward turn in the way we think about ourselves.[10] In search of certainty about his own existence, Descartes declared that the act of doubting was proof that we exist. Descartes elevated our ability to think for ourselves and on our own to the highest possible status, providing us with certainty of our existence. 'I think, therefore I am' is, however, increasingly at odds with the world being created by the web. Descartes urged us to look inwards, whereas the web urges us to turn outwards in the search for ideas. Descartes argued that thinking was a largely individualistic activity, while the web makes it increasingly social. In this We-Think world, creativity is invariably a collaborative activity that thrives when people share and mix ideas, allowing them to cross-pollinate. For Descartes, thinking ordered ideas inside our heads. When We-Think takes hold, what matters is social organisation: how we publish, debate, test, refine and reject ideas so that we think together. In the 20th century we grew accustomed to the notion that ideas came from specially gifted people, working in special places: the writer in the garret, the artist in the studio, the boffin in the lab. Yet with I Love Bees and Wikipedia, ideas are emerging from a mass of creative interaction among a wide range of people who combine different but potentially complementary insights. Our capacity for collaborative creativity will become ever more powerful because the opportunities to engage with others in creative interaction are expanding. The generations who grow up with these ways of thinking will have as their motto 'We think, therefore we are'.[11]

Importantly, though, just as I Love Bees and Wikipedia alert us to the possibilities of We-Think they also warn us that it flourishes only in very finely balanced circumstances. People gathering on social-networking sites, downloading user-generated videos or spouting off into the blogsphere do not create anything resembling collective intelligence. More often than not they produce a deafening babble or a deadening consensus, vicious disagreement or resounding reinforcement of already entrenched positions. On the web people seem either to argue or to agree with one another; it is much rarer for them really *to think* together. When they do, a delicate mix of ingredients is required, as Wikipedia suggests.

These seemingly allow people to be organised without having an organisation – by which I mean a clear hierarchy, job titles, an HR department.

We-Think's organisational recipe rests on a balance of three ingredients: participation, recognition and collaboration.

All successful efforts at We-Think – this book will introduce several more – depend on making it easy for capable participants to contribute to a joint project, whether by making an edit to an encyclopaedia entry, providing the answer to a clue in a puzzle, spotting a bug in a program, or tagging a piece of information. We-Think depends on motivating a mass of able contributors to get involved in a joint undertaking. As we shall see, the currency that draws people to these communities – from mine engineers in Cornwall, to kids playing computer games and the world's leading geneticists – is recognition. We-Think communities provide their participants with what they most value: recognition for the worth of their contribution, the value of their ideas, the skills of their trade.

The mass of individuals' contributions needs to be organised, however, so they connect, combine and grow, to create something robust and reliable, like a software program, a shared virtual world or a scientific theory. This calls for a mechanism that permits collaboration, for sifting good ideas from bad, better theories from worse. Without effective self-governance idealistic web communities, like so many communes and co-operatives before them, will collapse into an avalanche of diverse perspectives, rants, lies, gossip, falsehoods, truths and hearsay.

It is also critical that the contributors do not immerse themselves so fully into the collective that they stop thinking individually. Wikipedia is not a cult. People do not have to read the collected works of Jimmy Wales and attend local cells to

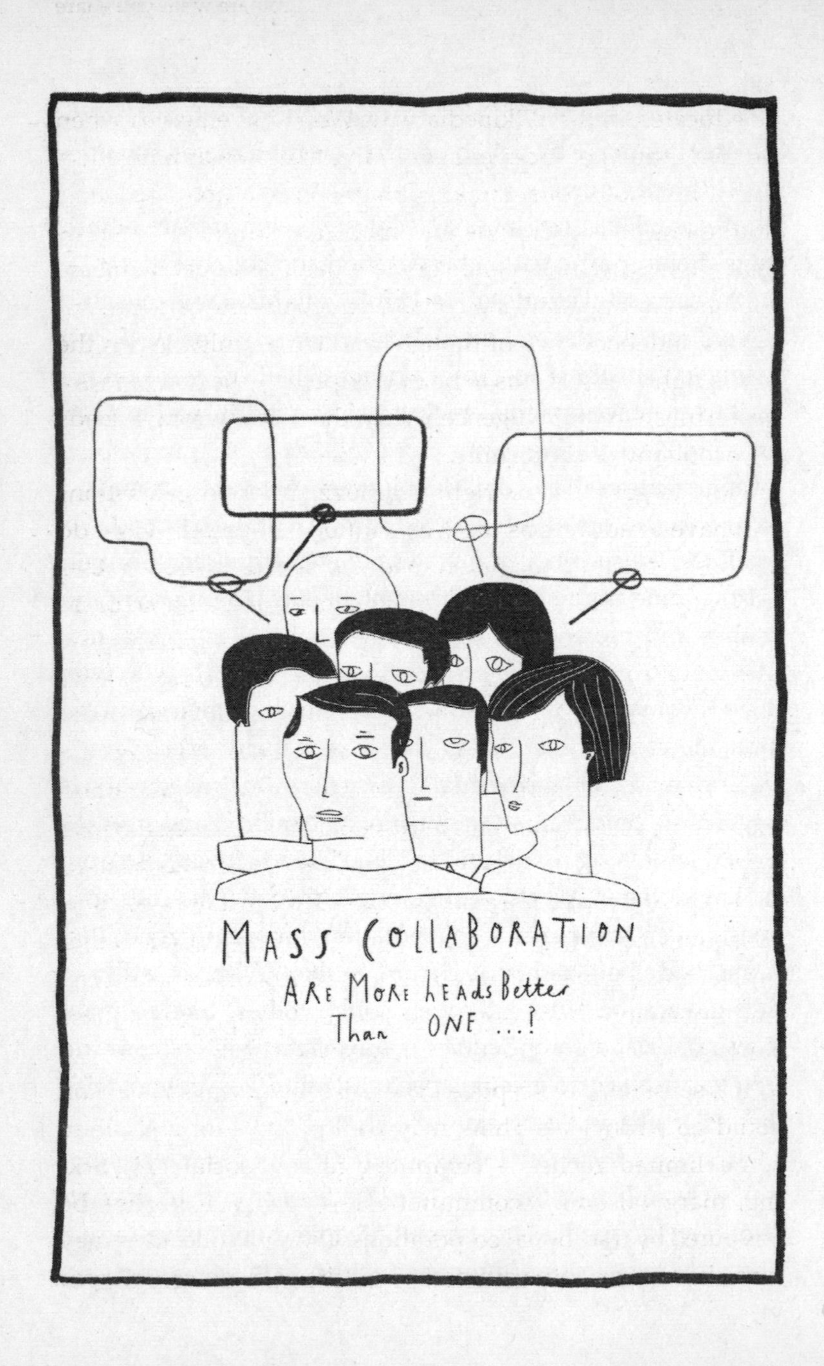
MASS COLLABORATION
ARE MORE HEADS BETTER
THAN ONE ...?

be educated in the Wikipedia way. We-Think emerges when diverse groups of independent individuals collaborate effectively. It is not group-think: submersion in a homogeneous, unthinking mass. Crowds and mobs are stupid as often as they are wise. It all depends on how the individual members combine participation and collaboration, diversity and shared values, independence of thought and community. When the mix is right – as it seems to be in Wikipedia – the outcome is a powerful shared intelligence. When the mix is wrong it leads to cacophony or conformity.

How to get that mix right is a puzzle more organisations will have to address as the web's influence spreads. How do all these contributions, often made by strangers, fit together to create a single working computer program, a game or an encyclopaedia? Why do masses of people work for free, first to create these things and then to give away the fruits of their work? In We-Think innovators share their ideas quite freely and welcome others' borrowing of their work and improving on it. They put a lot of unpaid effort into their innovations and then do not seek to profit from them. This is behaviour we have learned to regard as bizarre and yet on the web it seems to be part of the new normal. Can We-Think sustain itself, if its collectives do not earn any money to reinvest in their activities, let alone to pay the mortgages of their workers? And can traditional, top-down organisations find a way – given these constraints – to mobilise the power of We-Think?

It is sensible to be sceptical. There are many ways We-Think could go wrong. We-Think may well prosper for a while in some limited niches – computer games, social networking, marginal online communities – which will in time be devoured by traditional corporations. Or these collectives may turn themselves into commercial outfits, or perhaps collapse in on themselves, like so many failed Utopian communes in the past. The early experiments in We-Think we have seen so far could be no more than shooting stars, briefly lighting up the sky and distracting our attention before dying away, leaving everything much as it was. Certainly a great deal of the economy – chemical plants, railways, electricity generation, food production, banking, holidays – is not susceptible to this collaborative, open ethos.

My hunch, and the argument of this book, is that we are witnessing the birth of a different way of approaching how we organise ourselves, one that offers significant opportunities to improve how we work, consume and innovate. The logic of managerial capitalism is being scrambled up. To be organised we no longer always need an organisation, certainly not one with a formal hierarchy. Henry Ford's first mass-production factory emerged from a protracted period of experimentation at the start of the 20th century, when thousands of entrepreneurs were experimenting with small-scale ways to make cars. That experimental period may be where we are now. We may look back on the next 10 years as a period of immense opportunity to put in place a new way of organising ourselves, one that might have as much reach and impact as Ford's approach to mass production. We-Think could provide a different organisational base for society, one that encourages us to share more, to be more collaborative and participative, and in the process extends democracy, equality and freedom.

Ironically, as I shall argue in the next chapter, the success of We-Think will depend not on its being all new but on parts of it being quite old. The web is appealing in part because it offers to bring back to life more communal and collaborative ways of working which were sidelined by industrial organisations in the 20th century. The web's power comes from allowing us to be social in new ways. It speaks to a deep, old-fashioned yearning people have to be connected and to share – yet one that serves a modern purpose, to generate new ideas and knowledge. The oldest habits of sharing will be central to how we innovate together using new technologies. As innovation becomes more central to create less resource-intensive, environmentally damaging forms of economic activity, so will this ethic of sharing. As we will see, time and again, communities that share and develop ideas usually start around someone who donates their knowledge.

In 1672 Isaac Newton sent a long letter to Henry Oldenburg at London's Royal Society, outlining his theory of light and colour. Oldenburg printed the letter immediately in the society's *Philosophical Transactions* which he had created to provide for the fast and orderly dissemination of scientific discoveries. Newton's gift in making his ideas available for publication created a scientific community through which knowledge flowed for centuries to come. It was not a gift to a community that already existed. The gift created a community around it.

All the purest efforts at We-Think profiled in this book start with a gift of knowledge – whether that is software, tools, ideas or information – which then provides the basis for the growth of a community and the generation of yet more knowledge. Such communities allow commerce to thrive. But it is the communities that come first. Markets trade products; communities breed knowledge. Ideas do not live in the minds of individuals but through constant circulation as gifts. In the century to come, well-being will come to depend less on what we own and consume and more on what we can share with others and create together, especially as consumption becomes increasingly constrained by environmental concerns that mean we have to live more within collectively binding limits. In the 20th century we were identified by what we owned; in the 21st century we will also be defined by how we share and what we give away. That is why the web matters so much. It will allow us to share and so to be creative in new ways.

2
THE ROOTS OF WE-THINK

Imagine for a moment that a computer nerd, an academic, a hippie and a peasant get together for a joint project. The academic brings a belief that knowledge develops through a process of sharing ideas with colleagues and testing them through peer review. The hippie brings a deep scepticism about all sources of authority and a belief that an egalitarian community can organise itself. The peasant supplies habits that villages have long depended upon: the shared use of common resources like forests and fisheries, and a folk culture of stories and music which have been passed on by word of mouth. The geek offers to realise their dreams by networking them together with computers, modems and routers.

In a nutshell, those are the roots of the web-inflected culture we inhabit: a peculiar mixture of the academic, the hippie, the peasant and the geek. What binds them is a belief in the power of communities to share knowledge and other resources. Or to put it another way, the culture being created by the web is a potent mixture of post-industrial networks, the anti-industrial ideology of the counter-culture and the revival of pre-industrial ideas of organisation that were marginalised in the 20th century. Our expanding opportunities to be creative together come from this cocktail of ingredients. Let's start with the geeks and Web 2.0.

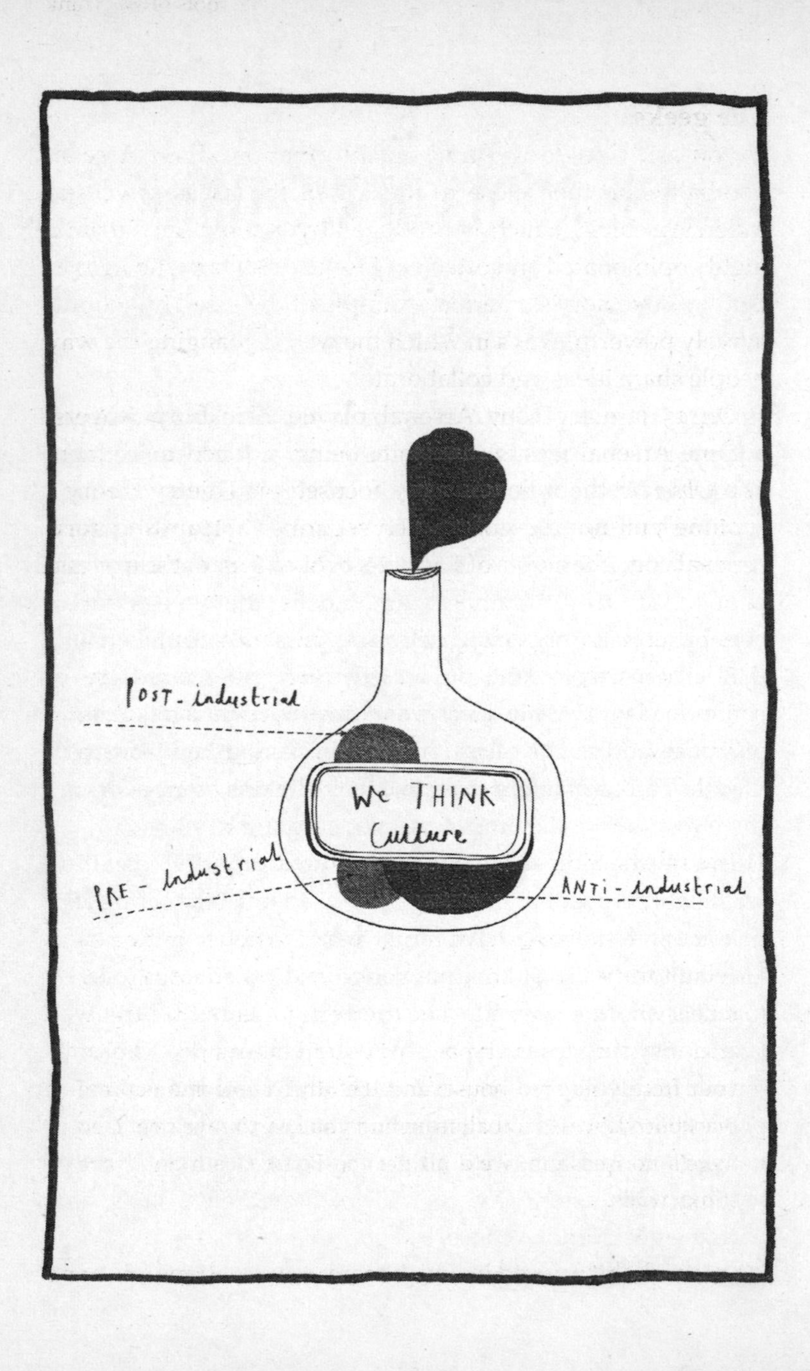
POST - industrial
We THINK
Culture
PRE - industrial
ANTi - industrial

The geeks

If you ask Google to find you information about Arsenal Football Club then close to the top of the list is a website called Arseblog, which is produced by a sometimes drunk, highly opinionated and often very funny Irish fan who lives in Dublin. Arseblog is a perfect example of the small but cumulatively powerful ways in which the web is changing the way people share ideas and collaborate.

On 13 January 2007 Arsenal played Blackburn Rovers, a game Arsenal won 2–0 despite being reduced to 10 men. The *Observer* the following day focused on Thierry Henry's sublime 71st-minute goal which secured his team's victory. Arsenal.com, the club's official site, evoked Roy of the Rovers:

> A performance brimming with class and character saw off Blackburn despite a red card for Gilberto after just 12 minutes at Ewood Park. Many teams would have buckled under those circumstances but Arsenal rolled up their sleeves, stayed true to their footballing principles and ran out worthy winners.

This was Arseblog's take on proceedings:

> If you were walking down the street and you saw two buses about to explode and one bus contained a squadron of killer robots who were going to give everyone on earth the plague, after they shagged everyone's wife and smeared poo all over your freshly painted house, and the other bus contained the Blackburn Rovers football team and you had to save one, then there's no question: we'd all get the Black Death and have stinky walls.

We now inhabit a world in which many organisations will have at least one independent full-time commentator like Arseblog agitating from the sidelines. Thanks to James Cherkoff of Collaborate Marketing for providing me with this story.

Arseblog does not just provide a daily outlet for its author's obsession. His daily posts provoke lively discussion, often attracting hundreds of contributions, and supply a round-up of the news in the online and offline editions of all the British newspapers, as well as from France and Spain whence many Arsenal players come. The site links to and often quotes from the other 15 decent blogs about Arsenal, as well as dipping into blogs run by fans of other clubs. When Alisher Usmanov, an Uzbek oligarch, became an unwelcome investor in the club, Arseblog was the best place to find out what was being said about him on the web, which was a lot more than in any newspaper.

Arseblog is a perfect example of how Web 2.0 is changing the way people relate to information and media. The web provides many more niches for people to start a conversation on something about which they feel passionately. The old, industrial media, newspapers and television, do not have enough room to cater for all the minority interests of their readers and listeners. Newspapers and television have high capital costs – studios and print plants – and to cover these costs they have to reach a large audience. The web with its much lower costs allows a committed and knowledgeable enthusiast to connect to his fellow fans. Web 2.0 differs from earlier, more static versions of the web, though, in that it encourages this community to have a conversation. The readers can post replies and talk among themselves. It is not one-way traffic. Arseblog is more than a niche publishing venture; it's the focal point for a conversation among a community of fans.

Tim Berners Lee probably was not anticipating Arseblog when he wrote the original software for the web at CERN, the European centre for nuclear physics research. Yet in some respects, it is a realisation of his original vision that the web could be a platform for collaboration, not just a new way to publish information. The now ubiquitous term Web 2.0 has spread like a rash since it was first popularised by technology commentator and publisher Tim O'Reilly in 2005.[1] By 2007, simply attaching the label 2.0 to something made it seem trendy, rather like the @ sign in the mid-1990s. Yet very few people seem to be clear what Web 2.0 really is. To avid web users the term is synonymous with a set of tools – like wikis and blogs – that allow people to publish and share information, including photos and videos. Software-programmers, meanwhile, focus on the underlying software that allows changes to appear on web pages without constantly refreshing them. Others use the 2.0 label to associate something with light-footed business models which are enabled by the web and which encourage and exploit user-generated content. None of these definitions gets to the heart of the matter.

The real reason Web 2.0 is so exciting is that it seems to promise a solution to a huge problem that confronts us: the unfathomably large quantities of information produced by a web in which millions of people are content-creators, not just receivers. In June 1993 there were 130 websites in the world. By mid-2007 there were 135 million registered host names and 61 million active sites.[2] Information is pouring in torrents from ever more sources as more people acquire the tools to become mini-publishers. The Pew Internet & American Life study in the US found that about 60 per cent of teenagers using the Internet regarded themselves as content-creators, using tools like blogs. In the UK a 2006 survey by Ofcom, the media regulator, found that 70 per cent of teenagers had created content online. That mass of content can make the web lush and attractive, but also like a sea of information that it is difficult to navigate.

Web 2.0 addresses this problem by capitalising on what is generated when millions of people share the same virtual space: their capacity for shared intelligence. Its tools allow us to collaborate to navigate our way through the jungle. The most obvious example is Google's search system, which works by treating a link from one website to another as a vote. If website A links to website B, that is like A voting for B. Websites that attract a lot of links also acquire more votes; so if website A itself has a lot of links to it then its vote for B is worth more than the vote of website C which has no links at all. When an Internet user starts a search, Google's algorithms count the number of votes people have made and make the votes a proxy for relevance and quality. When we ask Google to find something, it comes back with its account of our collective choices embedded in the links we have made: it is a collective intelligence service. As such it is one early, no doubt imperfect, answer to the web's cacophony and chaos. There are many others and they share the same recipe: we will not make sense of the mass of information generated by the web on our own; our only hope is to employ our shared intelligence. The more people contribute, the more we need to collaborate. The more people use the web to say 'I think ...' this, that and the other, the more we will need 'We think ...' to create some order, to sort the wheat from the chaff.

The history of blogging illustrates why we need We-Think to make sense of the mass of content being created. Blogging is a highly individualistic activity. What people blog about reflects their myriad interests and vantage points. The range of bloggers means it is impossible to create a centralised system of editing and quality control such as there is on a newspaper.

So the best way to find out which blogs are good is to rely on the judgements of other web users whom you trust.

The first blog was probably created in 1993 when the web browser company Mosaic set up a page on its website called 'What's New' and included links to other sites.[3] The writer Jorn Berger was the first to use the term 'weblog', in 1997; this turned into blogging in 1999 when Peter Merholz used the term 'wee blog' on his site, which quickly got shortened to blog. There were 23 weblogs in 1999,[4] and the first do-it-yourself blogging tool – Pitas.com – was created by Andrew Smales in July of that year to make it easier to create a blog, which by then had become an online diary.[5] Blogger.com launched a month later, became the most popular blogging software available and was subsequently bought by Google for an undisclosed sum. Blogger.com allowed many more people to become writers and publishers.

As more blogs were created, so elements of We-Think started to emerge. Sites that aggregated and collated blogs were set up, such as Brigitte Eaton's Eatonweb, which began with 50 blogs in early 1999; by 2007 it had 65,000.[6] That year Technorati, the specialist blog search service, claimed to be tracking 1.6 million updates a day to 75 million blogs, and monitoring the 175,000 new blogs that were being created every day.[7] Further sites such as Slashdot and Digg, Plastic and Fark aggregate and sort blogs and user contributions using tools like collaborative filters, while Trackback services allow bloggers to keep track of other blogs that are linked to their own.[8] The net result is that bloggers can swarm together and produce something like shared intelligence. In 2004 a self-organised investigation by US bloggers forced a television news network to withdraw a story claiming that President George W. Bush had received preferential treatment during his military service, by showing that the documents the network had relied upon were fakes. The website Slashdot, a meeting-place for nerds and geeks, gets 3 million visitors a day, mainly people who take part in scores of self-moderated discussions.[9] OhmyNews in South Korea brings together 55,000 citizen journalists to provide a news service that rivals that of traditional, mainly conservative newspapers and television stations.[10] YouTube and Flickr have enabled the widespread sharing of video and photographs and allow people to rate and sort content using tags and collaborative filters. So We-Think is vital to allow us to make the most of the extraordinary opportunities now available to us to create and contribute content.

Further, more highly developed collaboration can be found on social-networking sites like MySpace, Facebook and Bebo that make it easy for people to network around shared interests. Building on online bulletin boards created in the 1990s and on services like Friends Reunited – which linked long-lost school chums – the first social-networking site, SixDegrees.com, was launched in 1997. If proof were needed that an entrepreneur can be too early as well as too late, SixDegrees.com ended in failure. The next significant site, Ryze.com, launched five years later, fared better. Ryze focused on business networks and attracted about 250,000 regular users. Friendster, which started as a dating service for 20–30-year-olds in the autumn of that year, was the first to reach a significant audience, initially in Silicon Valley and then in San Francisco and New York. Users created profile pages with a variety of personal information and used these to link to others in a format that would become standard. Friendster had 3.3 million users by October 2003 but was subsequently overtaken by MySpace, with 78 million members in 2007, and in South Korea by Cyworld with 15 million members. By late 2007, however, the fastest-growing network was on Facebook, a site created by Harvard student Mark Zuckerberg to link to people in different residential houses on campus.

There are many more niche social networks, which may be the wave of the future, such as Gaia Online, an ecological network that claims 5 million members who have generated 850 million posts. Linked.In is business-focused: its 7.5 million members use it mainly for professional networking. Playahead, targeted at Swedish teenagers, has 1 million members. By 2007 all the US presidential candidates had profiles on social networking sites. In the Philippines, South Korea and Spain, the outcomes of general elections have been swung by groups organising themselves through online social networks allied to mobile phones. The most sophisticated of these – Nosamo, the online social support group for South Korea's President Roh – has established a structure for debating policy and making decisions in online forums.[11]

Social-networking sites work when they foster a spirit of collaborative self-governance: Friendster suffered a catastrophic loss of members through heavy-handed top-down management. Social-networking sites do not produce collective intelligence. But they create some of the pre-conditions for it by connecting together large groups of people with shared interests. Blogs and other tools allow people to contribute. Social networks allow them to connect. Still other tools are needed, however, for sustained creative collaboration to take off. The most famous is the wiki.[12]

Wiki is Hawaiian for *quick* but it is also an acronym for 'what i know is'. Wikis use a stripped-down version of the web's hypertext markup language which is why they were first known as the 'quick-web'. The first wiki was created in March 1995 by Ward Cunningham, a prolific programmer based on

Portland, Oregon; by its 10th anniversary Cunningham's wiki had 30,000 pages.[13] A wiki allows anyone with access rights to log in and directly edit what is on the page. This process can be extremely open – as it it with Wikipedia – or confined to groups within an organisation. As well as Wikipedia and its offshoots there are numerous hosted wiki services – known as wiki farms – such as SocialText, EditMe, OpenWiki and Swiki. Wikis allow people to co-operate to summarise a debate or amass a body of information and create documents with a collective author. They work best when many minds can focus on a shared task with some clear goals: to create an encyclopaedia, plan a meeting, update a list, assemble scientific data or write a report. They are less effective in reconciling competing opinions. Joe Kraus, co-founder of JotSpot, the wiki software company, says most wikis are best suited for small, well-defined groups of people collaborating on projects of limited duration.[14]

The web will work for us best when the power of mass collaboration orders the chaos of mass self-expression. A string of tools – laptop computers, mobile phones with digital cameras, blogging software – has made it easier for people to contribute their views and ideas through blogs, photographs and videos. But on their own these just create a morass of information, which it is difficult to sort through and navigate. There is more 'I think ...' than ever. That makes the web rich but messy. The best way through this plethora of material, to find the person and the information you are looking for, is to rely on We-Think: the power of shared intelligence to sort wheat from chaff, whether through search engines, collaborative filters, wikis, or recommendations from trusted blogs and friends on social networks.[15]

The web's tendency to promote sharing is not just a matter

IDEA

of necessity and some clever software. It has deeper roots in the academic and hippie culture that spawned the web in the 1960s. The best way to understand that story is to start with Doug Engelbart.

The digital communards

Nearly 40 years ago, Doug Engelbart stood before the cream of America's fast-emerging computer industry gathered in Brooks Hall in San Francisco. On a giant screen he showed how they might: edit text directly on the screen of a computer using a keyboard and a mouse; insert links into a document leading to another related document; mix text with graphics and video; and connect one computer to another several miles away using a phone line. He argued that these developments would create the basis for a computer-to-computer network that would allow people to work on the same documents even when they were on separate continents. Engelbart was showing his audience the bare bones of what would become the Internet. It was 9 December 1968.

To many in the hall, brought up with computers the size of rooms that used punch-cards to complete complex calculations, it must have sounded crazy. Yet in just 90 minutes, Engelbart showed them what would become the future of computing and communications and would come to underpin modern commerce and culture. As Fred Turner puts it in his history of the period,

> For the first time they could see a highly individualised, highly interactive computing system built not around the crunching of numbers but around the circulation of information and the building of a workplace community.[16]

From then on, it was possible to see computers as the bearers of social and organisational revolution.

Engelbart had joined the Stanford Research Institute in Palo Alto, California, in 1957 and three years later wrote a memo for the US air force arguing that something akin to personal computers could be yoked together with 'associative links' in documents, a precursor of the web's hypertext. By 1963 Engelbart had set up a research centre to study the augmentation of human intelligence – and between 1966 and 1968 he and his team created the Online System, dubbed NLS, which provided key ingredients for what became the Internet. By 1974 Engelbart's lab was riven by disputes and closed down. The computer industry did not realise his dream until cheap personal computers and the Internet were combined in the mid-1990s. Before Engelbart, the computer was distrusted as a dehumanising tool of corporate and bureaucratic control. His work re-imagined it as an instrument of personal liberation and freedom of expression, with the potential to flatten hierarchies, decentralise organisations and unleash collective creativity.

The man recording Engelbart's presentation in December 1968 was Stewart Brand, a 29-year-old itinerant artist and journalist. Brand's eclectic interests meant he had links with avant-garde artists in Manhattan who were exploring new art forms; with backwoods communes in New Mexico where people were exploring new ways of living; and with the counter-culture of San Francisco, where technology, protest and drugs fused together. As technologists like Engelbart were imagining new ways for people to collaborate using computers, others were experimenting directly with communal living: by 1970 about 750,000 people were living in tens of thousands of recently established communes, in search of a simpler, more authentic way of life. Brand stood at the crossroads between bohemianism and new technology, the original digital communard.

Brand's most significant contribution was the creation in 1968 of the *Whole Earth Catalog*, a mixture of news, tools, reading suggestions and mail-order offers of everything from tantric art to cybernetics. The first rough-and-ready version of the *Catalog* sold 1,000 copies. By the time it closed three years later it had sold 1.5 million, and Brand won a National Book Award for his efforts. The last copy had 448 pages, listing 1,072 interesting items. The *Whole Earth Catalog* contained elements now recognisable in trendy Web 2.0-style businesses like eBay and Craigslist. Much of the content was submitted by readers, and those who were first to recommend something interesting got their names listed in the magazine. Brand went on to help create the Whole Earth 'Lectronic Link, an early Internet bulletin board, which in turn spawned the Electronic Frontier Foundation, which campaigns for freedom of speech online, and *Wired* magazine, the bible of the New Economy. More than any other magazine, *Wired* lionised technology entrepreneurs as the carriers of revolution.

By 1971, however, the workload on the *Whole Earth Catalog* was taking its toll on Brand and he decided to close the magazine down with a Demise Party, held on 21 June at the Palace of Fine Arts in the centre of San Francisco. The entertainment included clowns, belly dancers, trampolinists and a band called Golden Toad who played Irish jigs and Tibetan temple music. Brand, dressed in a monk's black habit, brought with him $20,000, the sum he had invested in starting the *Catalog*. At 9.30 p.m. he announced that anyone could take the stage and suggest a scheme the money should be spent on that would keep the *Catalog*'s spirit alive. Brand wrote down suggestions from about 50 people on a blackboard. There was little agreement. Quite a bit of the money disappeared into the throng. But by the end of the night, when most of it had been retrieved, the man standing at the microphone was Fred Moore; he had $14,905 in his hands. He turned out to be one of the patron saints of the collaborative web.

Fred Moore, the son of a military man, decided in his teens that he was a pacifist. In 1959 when he arrived at Berkeley to study science, he looked no different from the average American teenager: white socks, braces, tennis shoes and neatly turned-up jeans. It had been compulsory at Berkeley since 1868 for male students to undertake military training. Moore set up a table during freshers' week soliciting support for a campaign against compulsory military training. He was summoned to the Dean's office and told exemptions would be given only on grounds of physical disability, previous military experience or foreign citizenship. On the morning of 19 October 1959 Moore sat down on the steps of Sproul Hall, the university's administration building, with a canvas mat, a pint of water and a petition, and started a seven-day fast against compulsory military training. Moore was the original student protester. Thanks to him, military training was made optional. In the 1960s, the decade of civil rights and the Vietnam War, hundreds of thousands of students followed his lead and turned campuses into places of protest.

Moore was an odd person to have been left holding the money at the end of the Demise Party, because he regarded money as evil. Yet the cash helped to keep him going as his interest shifted from protests to computers as a tool for social change. In the early 1970s, Moore got involved in the People's Computer Company, and in 1975, when that started to split apart, he and fellow volunteer Gordon French launched a club for amateurs interested in the social impact of computers. Moore cycled around Palo Alto posting flyers which announced:

> Amateur Computer Users' Group, Homebrew Computer Club … you name it. Are you building your own computer … if so you might like to come to a gathering of people with like-minded interests, exchange information, swap ideas, talk shop, help work on a project, whatever …

The club's first meeting, in French's garage, attracted 32 people, six of whom had built their own computers. The club embodied the hacker ethic: people making things for themselves and helping one another to do the same. Twenty-three high-tech companies can trace their lineage to this club of do-it-yourself amateurs, among them Apple. Fred Moore died in a car crash in 1997 at the age of 55, largely unknown, and yet he helped to shape modern America: the original student protester and co-founder of the club that spawned much of the digital revolution we now live with.

After the Homebrew Computer Club's first few meetings Lee Felsenstein, a computer engineer who had also been involved in student protests against the Vietnam War, became its chair. Felsenstein wrote for an avant garde publication call *The Tribe* and in 1973 he led a project called Community Memory, which consisted of two computer terminals set up in a popular record shop and the university library, linked to a central computer.[17] Community Memory was an early prototype for what Web 2.0 would become. The project described itself as

> An actively open information system, enabling direct communication among its users with no centralised editing or control

> over the information exchanged. Such a system represents a precise antithesis to the dominant uses of electronic media which broadcast centrally-determined messages to mass passive audiences.

Felsenstein had developed this convivial approach to the use of computers as a tool for everyday self-expression and collaboration after reading the work of the radical philosopher Ivan Illich, whose ideas provided the backdrop to much of the discussion among the high-tech bohemians of San Francisco. People like Felsenstein, Moore and Brand turned to computers in part to realise Illich's ideas.

Illich spent his life transgressing boundaries and counfounding conventional wisdom. Trained as a priest and rapidly promoted in the Catholic hierarchy, he became a fierce critic of the Vatican. For much of the 1970s he was a darling of the left, sharing common ground with Herbert Marcuse in his critique of a one-dimensional society run by large corporations. He was an environmentalist before the movement had a name. Yet Illich was also a libertarian who dismayed many of his left-wing fans with a withering attack on Castro's Cuba and enraged feminists with his defence of traditional gender roles.

Illich was born in Vienna in 1926, the son a civil engineer, and grew up in a comfortable middle-class home. The Nazis expelled him from Austria in 1941 because of his mother's Jewish ancestry, and for the remainder of his life he was a wandering intellectual living with few material possessions.[18] In a golden period in the 1970s Illich set about dissecting the failings of modern institutions, the professionals who organise them and the systems they design, in a series of short polemics: *Deschooling Society, Limits to Medicine, Disabling Professions* and *Tools for Conviviality.* He argued that as people become dependent on the expert knowledge of professionals they lose faith in their own capacity to act. His solution was that people should spend less time as consumers, more as producers of their own well-being. And for that to be possible they need more convivial, easy-to-use tools.

Illich's most optimistic book, *Tools for Conviviality*, which inspired Felsenstein and others in the hacker community in the 1970s, put the challenge this way:

> I believe a desirable future depends on our deliberately choosing a life of action over a life of consumption, on our engendering a lifestyle which will enable us to be spontaneous, independent, yet related to each other, rather than maintaining a lifestyle which only allows us to produce and consume.

Convivial institutions work through conversation rather than instruction; through co-creation between users and producers, learners and teachers, rather than delivery from professionals to clients; and through mutual support among peers as much as by means of professional service. In *Deschooling Society*, published in 1971, Illich provided some principles for how a more convivial education system might work. These included providing access to resources for learning at any time, in airports, factories, offices, museums and libraries as well as in schools; enabling those who want to share knowledge to connect with those who want to learn from them, through skills exchanges and directories of classes from which people could choose; and allowing those who want to propose an issue for discussion and learning to do so easily. In 1971, all that must have sounded mad. In the era of eBay and MySpace it sounds like self-evident wisdom. The collective self-help of We-Think is an attempt to realise some of Illich's ideals.

Illich was not the only philosopher who provided ideas to shape the way technologies might be used. E. F. Schumacher, the author of *Small Is Beautiful*, argued for a society of 'production by the masses not for the masses'. Marshall McLuhan in *Understanding the Media* extolled a pre-bureaucratic humanism based on a retribalisation of society. In 1968 Roland Barthes, the French literary critic, announced that only the 'death of the author' as the sole arbiter of the meaning of a work of art would clear the way for the 'birth of the reader' as a participant, actively engaged in making sense of a text. Guy Debord, the founder of the anarchist group Situationist International, turned that idea into a manifesto. Debord denounced modern society as no more than a 'society of spectacle': 'The spectacle is the opposite of dialogue. It is the sun that never sets on the empire of modern passivity.' The avant-garde imagined that spectatorship would give way to participation permitting people to become more social and collaborative, egalitarian and engaged with one another, to borrow and share ideas.

Our ever-so-trendy Web 2.0 culture, embraced by politicians of left and right, by companies and educators, is the bastard offspring of this *mélange* of ideas from the 1960s counterculture. Mass participation, Debord's antidote to the society of the spectacle, has turned into YouTube and social-networking sites on which we can all make a spectacle of ourselves. The libertarian, voluntaristic communities that briefly flowered in their thousands in California and New Mexico find their modern counterparts in the open-source communities listed on SourceForge and the virtual homesteaders of Second Life, as they create their own rules and currencies. Collage and pastiche, recombining ingredients provided by others, were central not just to Situationism but to futurism, cubism, Dadaism and pop art. The rip, mix, burn generation of Apple iPods, hip-hop music and YouTube videos is Debord's heir. The We-Think generation is living out the hopes of the 1960s radicals for the creation of a harmonious, post-scarcity society that is free, decentralised and yet apparently egalitarian, a world in which as Fred Turner put it 'each individual could act in his or her own interest and at the same time produce a unified social sphere, in which we were "all one"'.[19]

As an offspring of the 1960s the web also carries many of the congenital weaknesses that afflicted the counter-culture. The communes collapsed because they could not govern and work efficiently. They claimed to be egalitarian and open and yet they were mostly for the white, college-educated middle classes. Women were largely confined to domestic tasks. Even more significantly the Homebrew Computer Club bequeathed to us an argument that has since become the longest-running dispute of modern times. That dispute pitted the ethics of mild-mannered Fred Moore against those of Bill Gates, the economics of sharing against the economics of private ownership.

At the Homebrew Computer Club's third meeting in 1975 someone 'borrowed' a copy of a software program called Altair BASIC, designed to run the first personal computer, the Altair. The program, on a long roll of paper tape, was the first made by a tiny Albuquerque company called Micro-soft, created by William Gates and Paul Allen. The tape fell into the hands of Dan Sokol, a semi-conductor engineer who had access to a high-speed copying machine. BASIC software was freely circulating in academic circles, and Sokol could not see why personal-computer users were being asked to stump up $500 for a barely modified version. At the club's next meeting he handed out 70 copies of the Microsoft program. Gates was incensed that his tiny company was being threatened by the amateurs. In an infamous letter to the hobbyists, Gates complained,

> As the majority of hobbyists must be aware, most of you steal software. Hardware must be paid for, but software is something to share. Who cares if the people who worked on it get paid?

That started the digital civil war over how information should be owned and paid for. What began as a row in a tiny club of enthusiasts in Palo Alto has spread into hundreds of millions of living rooms, bedrooms and increasingly courtrooms around the world and involves not just software but film, music and virtually every other kind of information as well. Fred Moore's co-operative ethic – the more you share, the more value gets created – ran straight into Bill Gates's hard-headed economics – you profit from what you own. When Microsoft launched its Vista operating system in 2007 with security features to allow the company to keep tabs on how it was used, it was firing the latest salvo in this civil war.[20] Like most civil wars this one is messy and complex, because people seem to swap sides. Big companies like IBM and Hewlett Packard make money from implementing the open-source software programs that they help to create. Google earns vast sums by milking the web's collective intelligence: never has so much money been made by so few from the selfless, co-operative activities of so many.

The web started life in odd circumstances for an innovation that will change the world. It did not come out of commerce or a research-and-development laboratory but can instead trace its roots back to a mix of academics, visionaries, hackers and hippies who all believed in, and often worked in, non-market communities of equals, in which designers of computers and software were also the users. The pioneers of open-source programming and online communities in the 1960s and 1970s talked the language of fellowship and communion.[21] The web has never shaken off these communal and collaborative roots, which is why money-making companies have found it hard to bend it to their commercial purposes. At best they cohabit, often uneasily. The nature of that cohabitation between commerce and community, what we own and what we share, will shape much of the future in science, culture, politics and economic life.

The developing world could be far more adept at this cohabitation than the rich developed world. This is because contemporary web culture is reviving older, pre-industrial forms of organisation that in the developed world were sidelined by industrialisation. In much of the developing world these pre-industrial ideas are alive and well. What will happen when the networks created by the geeks combine with the traditions and habits of millions of people who were until recently rural peasants?

Radically old ideas

Ivan Illich was ahead of his time by being a long way behind the times. His critique of industrialisation harked back to pre-industrial, more communal and less hierarchical forms of organisation, in which local, low-technology production was best equipped to meet demand. The web is reviving older ways of organising that were cast to the margins during the industrial era. The new turns out to be a lot older than we thought.

We-Think revives the idea that sharing and mutuality can be as effective a base for productive activity as private ownership. It draws on the tradition long established in villages and communities to use common pools of resources. A commons is anything like the streets on which we drive, the skies through which planes fly, public parks and the beaches on which we relax.[22] A commons belongs to a community – sometimes a tightly defined community, sometimes everyone – and is usually governed by common consent of the people using it, whether Spanish farmers watering their oranges from shared irrigation systems or Turkish fishermen taking turns over which areas they can fish. A day on a decent public beach exemplifies the commons in action. Beaches are ordered without being controlled. No one is in charge. A public beach is a model civic space: tolerant, playful, self-regulating, democratic in spirit. As the day unfolds everyone takes their spot, adjusting to where everyone else has pitched their towel. There are no zoning regulations, fences or white lines to tell you where to go (admittedly this is not true of some beaches in Spain, France and Italy). Order emerges as each family joins the throng. On popular beaches people spend all day in close proximity but generally remain civil and considerate. Precisely because there is no one in control – other than sometimes lifeguards looking after safety – people take it upon themselves to self-regulate. Beaches are generally egalitarian in spirit because barriers to entry are quite low. Normal rules do not apply because there is no private property. A public beach is a commons for pleasure. Many public events and spaces – festivals and carnivals, parks and libraries – thrive on this ethic of mass self-regulation. The web is bringing the spirit of the beach into the sharing of ideas and information.

The beach also explains why the enclosure of the cultural

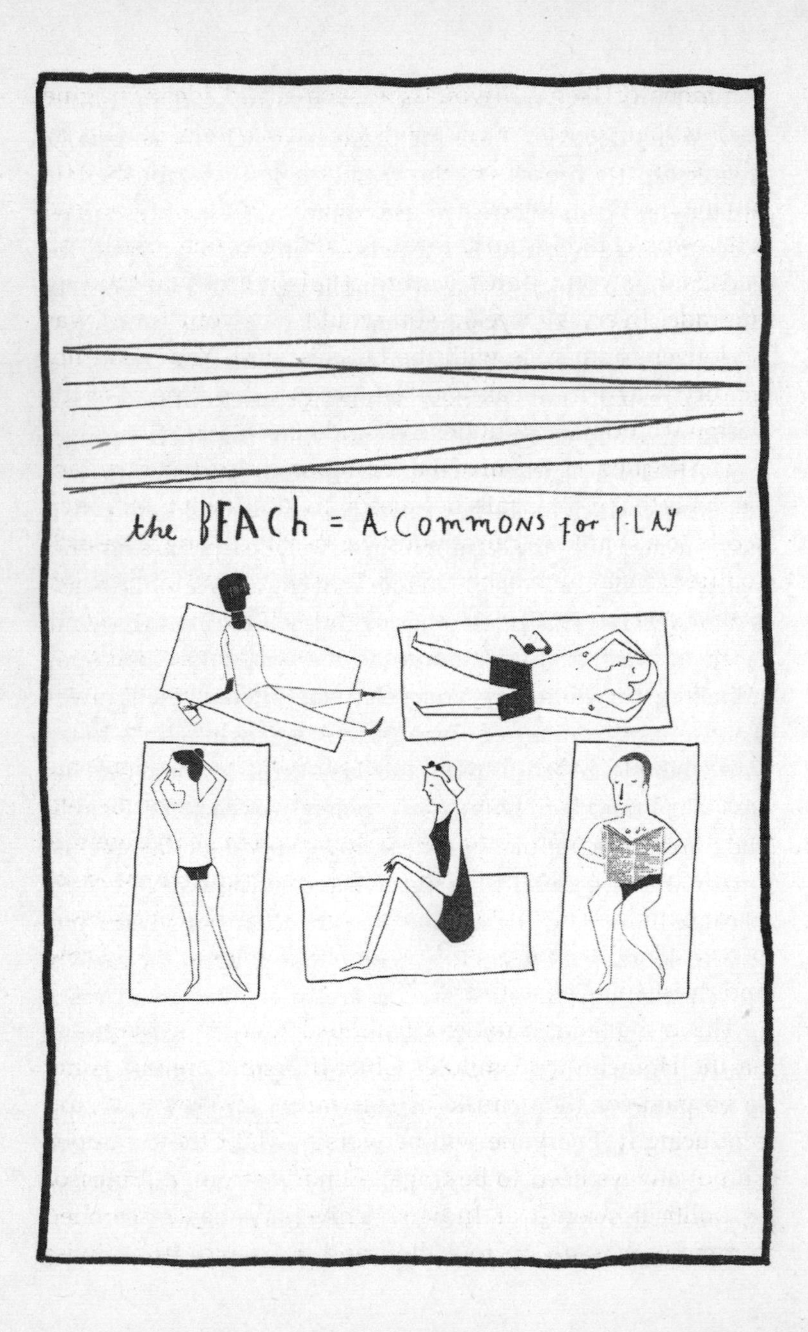
the BEACh = A COMMONS for PLAY

commons of Web 2.0 would be such a bad idea. Imagine finding that your favourite public beach had been bought by Microsoft. You would be able to get on to the beach only by buying Microsoft towels and windbreaks. You would be told where you could lay your towel according to how much you had paid. If you wanted to surf, it might cost you more to upgrade. Every two years you would find your towel was no longer compatible with the beach's sand. You could not modify your windbreak yourself, because key aspects of the design would be kept under lock and key.

The frequent criticism of the commons as a system of organisation is that it often fails because individuals with unlimited access to a shared resource will over-use it – taking too much cod from a common fishery, or over-grazing land. Something that is everyone's property fast becomes no one's property and so gets abused. As soon as that possibility rears its head, everyone using the commons sees no reason to restrain their use. It is argued that the only ways to avoid this 'tragedy of the commons'[23] are to fence the commons into pieces of private property that people look after or to place the whole thing into state ownership. This 'tragedy of the commons' argument is currently being used by media, music and film companies to fence round the cultural commons of the web. Open access is perceived as leading to abuses, such as rampant file-sharing and the stealing of software.

This is the same argument that Bill Gates used in his attack on the Homebrew Computer Club: if people are not going to be paid for their music or software then they will stop producing it. Everyone will be worse off. Yet the commons do not always need to be tragic. Elinor Ostrom, a professor of political science at Indiana University, has researched commons that are up to a thousand years old, from Swiss villages that share summer grazing to rice-paddy irrigation systems in the Philippines.[24] Ostrom found that the commons can work when they are self-governing, when participants can easily monitor one another's behaviour and when sanctions for breaking rules are effective because people want to be part of a community in which reputations matter. She suggests a commons works best with a bounded community, but even large and open communities can sustain a commons if they are broken down into smaller platoons or guilds. A commonly owned irrigation system can be the shared platform for a mass of orange farmers. Commerce has often been built on shared foundations.

This is especially so when the commons are not finite resources like woods and land, but knowledge and ideas.[25] Bodies of ancient music are part of a cultural commons. Einstein's theory of relativity is part of the commons, along with Watson and Crick's description of the double helix of DNA. Language is a kind of commons. Dictionary publishers, broadcasters and teachers may attempt to claim authority over language but their right to do so is frequently contested. A language is not owned by anyone. The English language, which has developed by absorbing so many foreign influences, now provides a global common resource for business, science and culture.[26] English has an estimated 380 million core speakers. More than 300 million have English as a second or third language. English is thought to be the lingua franca of more than a billion people – and rising. A key feature of the English language has been the ease with which it has borrowed, stolen and adapted words from other languages. By the end of the 16th century, it had incorporated words from around 50 different languages. The idea of a private language spoken only by a single individual is an absurdity. Language lives in conversation between people, and its structure is held in place by public rules. And a language grows with use rather than wearing out.[27] Sheep grazing the same field will soon eat up the grass. On a cultural commons, such as a language, encyclopaedia or game, the grazing sheep still eat up the grass, but because they excrete more the more they eat, the field – the commons – regenerates itself and grows fresh grass.[28]

Organisations that mobilise peer-to-peer ways of working are the height of fashion. Social-networking sites like Facebook rely on people making connections with one another through lateral networks of friends and contacts. Most open-source projects work only because there is rigorous peer review of software code produced. Yet although the scale of the recent explosion of peer-to-peer activity is new, the idea itself dates from the late 16th century.

Until the mid-16th century, most science was conducted under conditions of secrecy. Science was likened to alchemy and sorcery. By the second half of the 17th century, learned scientific societies were flourishing across Europe, embodying the practices of the modern web – the open disclosure of, and debate about, ideas. According to Paul David, a professor of economics at Stanford University, the system of peer review has noble roots.[29] Kings and princes surrounded themselves with poets and musicians, architects and scientists, to provide entertainment but also to project their prestige. Yet science posed a tricky problem. As science became more theoretical and mathematical, patrons found it harder to judge for themselves how good their scientists were. Communities of scientists emerged, David argues, to allow scientists to establish their credentials. The idea that your peers are the best judge of how good you are – especially at something difficult to measure such as intellectual work – started with Henry

Oldenburg's Royal Society, founded in 1660. In the following decades scores of scientific and learned societies were created across Europe, in order that scientists could earn a reputation among their peers.

A social innovation created in pre-capitalist Europe in the 16th and 17th centuries for an élite of nobles and scientists is becoming a mass way of working thanks to the web. When people say there is something oddly noble about the way open-source programmers give away the fruits of their labour without being paid, they are closer to the truth than they might think. Gaining recognition from your colleagues through peer review stems from an aristocratic tradition of science in which money really did not matter.

Common platforms and peer-to-peer working allow innovation to emerge from a community. Communities of innovation are all the rage on the web but, once again, they are very old. Community and conversation are the roots of creativity. Ideas live within communities as much as they do in the heads of individuals, as shown for example in the 18th-century Cornish tin-mining industry just before the industrial revolution.

Cornwall was the Silicon Valley of its day, home to the most impressive innovations in industrial technology. Cornish tin and copper mines posed the trickiest problems for engineers and so demanded the greatest ingenuity. The deeper the mines went, the more prone to flooding they became. In 1769 the inventor James Watt came up with an engine design that incorporated a separate condenser, which cut the amount of coal needed by two-thirds. This transformed the economics of mining. The Watt engine, which he marketed with his business partner Matthew Boulton, quickly spread through Cornish mines – but the mine-owners became disenchanted.[30] Boulton and Watt charged them a royalty fee equivalent to a third of the amount of money that a mine saved each year after the installation of their engine, the design of which was protected by a very broad patent enforced ferociously. Mine-owners soon started to complain. The patent meant they could not improve on the design through their own efforts. Boulton and Watt had no incentive to make further improvements because they were making so such money. In 1790 Cornish mine-owners revolted. Today they would be denounced as software pirates: they started to install unauthorised versions of the engine. Boulton and Watt took them to court and got their patent's lifespan extended until the year 1800. Innovation ground to a halt. Boulton and Watt never made another sale in Cornwall. In 1811, a group of mine captains – lead by Joel Lean, a respected local engineer – founded a journal to share new ideas, in the spirit of collaboration and open competition that often marks creative communities. *Lean's Engine Reporter* was published each month for almost a century, reporting on all aspects of engine design. A year after the *Reporter* started, Richard Trevithick and Arthur Woolf introduced a new design that fast became the industry's standard operating system. Woolf and Trevithick did not patent their design and freely allowed other mines to copy from the original erected at the Wheal Prosper mine. They made their money installing, adapting and improving engines. The tightly knit community of Cornish engineers were soon swapping ideas through the *Reporter* about how to improve on the basic design.

During Boulton and Watt's ascendancy, following an initial leap, innovation stalled. The open and collaborative period that followed produced near continuous innovation for more than 30 years, as a host of practitioner-engineers improved upon Woolf and Trevithick's design. None of this innovation was patented. By 1845, engines in Cornish mines were more than three times more efficient than the Boulton and Watt engine of 1800. They became known as 'Cornish' engines in recognition of the cumulative, collaborative and collective nature of the innovation. During this period Cornwall had the fastest rate of steam-engine innovation in the world and the lowest rate of patenting in Great Britain.

The Cornish engine story prefigures today's contest between Microsoft and open-source software: sharing can be a highly effective basis for commercial endeavour. In Cornwall rival firms released to one another ideas that brought significant cost reductions to all. They did so because the mine-owners had a strong shared interest and independent mine engineers were keen to make known what they had achieved. The Cornish tin mines ran on open-source software centuries before the computer. We are now seeing a revival of this communal approach to innovation, its tool being the web rather than *Lean's Engine Reporter*.

The social approach to creativity encouraged by the web is reviving one of creativity's oldest forms – folk – by empowering a mass of amateurs to create and share content. In 2006 and 2007 do-it-yourself, user-generated content became hugely fashionable. Many of the advertisements shown during the 2007 Superbowl were made or inspired by amateurs. Lasse Gjertsen, a 22-year-old from Larvik in Norway, became an international star by making intricately cut home videos that attracted an audience of 2 million on YouTube. In 2007 *Time* magazine anointed us all Person of the Year to mark this surge in collaborative, everyday creativity. Yet once again this is simply bringing back to life an older, folk culture, which was extinguished by the mass-produced, industrial culture of the record and film industry of the 20th century.

Over the next few years we are likely to witness the growth

of an enormous, collaborative, digitally enabled vernacular culture that will be both more democratic and creative than what preceded it but also more raucous and out of control. Many of the themes of folk culture recur in discussions about the growth of social networking, blogging and YouTube. Folk has always rested on a cult of authenticity: the self-taught singer-songwriter armed with just an acoustic guitar; the self-taught artist making a sculpture from driftwood. Now we have legions of self-taught amateur musicians, armed with GarageBand, posting their work on MySpace, their videos on YouTube. Nothing is fresher in our manufactured and commercialised culture than raw, untutored talent, untouched by training and the lure of commerce.

Folk has always been art for people outside, and at odds with, mainstream culture and commerce. Folk artists are in it not for the fame, but for the pleasure of producing good art, using an everyday style rather than the formal, self-conscious and fancy styles taught in art and music colleges. By making tools of cultural production ever more widely available, Web 2.0 has unleashed new waves of authentic talent – pensioners on YouTube, bloggers like Salam Pax, or Internet performers like Ze Frank and Ask a Ninja, who can find audiences without succumbing to the cookie-cutting marketing of the mainstream culture industry. The wave of digitally enabled folk culture is presented as an antidote to the plastic, celebrity-obsessed and contrived culture of mainstream television with its heavy reliance on reality TV formats. The more shaky, grainy and real video looks, the more real it must be; professional advertisers and agencies crave the raw and fresh.

Folk rests on a self-styled communal creative culture – songs and tunes borrowed and passed on for generations. Claims to sole authorship or celebrity are frowned upon. The point is to rework material created by others, to pay homage to those who went before. Authorship is lost in the mists of collective creativity. Folk artists encourage others to borrow from them rather than protecting their rights as authors. As Woody Guthrie's copyright notice put it:

> This song is Copyrighted in U.S., under Seal of Copyright 154085, for a period of 28 years, and anybody caught singin it without our permission, will be mighty good friends of ourn, cause we don't give a dern. Publish it. Write it. Sing it. Swing to it. Yodel it. We wrote it, that's all we wanted to do.

Guthrie's copyright notice could be a rallying cry for the file-sharing We-Think generation.

Dorothy Noyes, a professor at Michigan University and the leading folklorist in the US, points out that we have to go a long way back to understand the future: epic poems such as the *Iliad* and the *Odyssey* developed over many years through the contributions of probably several hundred poets and performers all over the Greek world. The *Iliad* and the *Odyssey* had a core code, which many people worked on, specialising in particular scenes or episodes, to improve it. There was no master text until much later. The Homer to whom the poems are attributed is, as Noyes notes,

> the name we give to what was a collective process of creativity involving many people over a long period. The founding myths of our culture set out in classical drama, drawn upon time and again by subsequent writers, were not created by a single author but a highly collaborative, social process of creativity.

Before the mass-produced book – that is for most of human history – most culture and art was folk. In many developing world societies it still is. In Ghana, for example, designs for traditional clothes are owned by communities that include craftsmen living and dead, royal patrons and entire villages. The same is true for Aboriginal stories. The popular culture of We-Think is the mutant offspring of the marriage of folk culture and digital technology.

We-Think culture is a hybrid of these odd ingredients: the geek, the academic, the hippie and the peasant. Media and culture used to be an industrial business dependent on large printing presses and expensive television studios, making products for the mass audiences needed to sustain their costly operations. The spread of the web means more people than ever can have their say, post their comment, make a video, show a picture, write a song. The more I-Think there is, the more content and information we create, the more we will need We-Think to sort it. The anti-industrial ideas of the 1960s counter-culture, bringing together Doug Engelbart's vision of distributed and decentralised technologies and Fred Moore's ethic of sharing, underpin hopes that we might still be able to create egalitarian, self-governing communities. Finally, We-Think revives pre-industrial forms of organisation: the commons, peer-to-peer working, community innovation and folk creativity. We-Think is so potent because it mixes the brand-new – the blog and the wiki – with the very old. The achievement will be to turn these basic ingredients into a working model, a social form of creativity in which many contributors have the capacity and tools to think, act and experiment together, that is both informal and structured. It is not enough for people merely to participate, to have their say; they have to find ways to collaborate and to build on what others are doing, so that whatever they are engaged in grows through accretion, mutual criticism, support and imitation. When people do pull off this trick – which is by no means always – they can create complex, valuable, reliable products: encyclopaedias, software programs, computer games, news reports, scientific theories, epic poems. How they do it and when they do it is the subject of the next chapter.

3
HOW WE-THINK WORKS (AND NOT)

Crowds are not automatically wise and mobs are not necessarily smart. It all depends on how they are made up and come together. Why do some collaborations turn into We-Think – seemingly generating a momentum and intelligence of its own – while others do not? Why do *any* of them work? Viewed through the lens of traditional industrial-era organisations, it seems improbable that skilled people, with busy lives, would give their time for free to mass collaborative efforts, only to give away the fruits of their labour. Or that their many contributions could be brought together into a coherent whole, rather than descending into chaos, fragmenting or getting bogged down in committees and debates about what to do next. So how do creative communities avoid becoming inward-looking cliques that ignore new ideas brought by outsiders?

Successful attempts at We-Think answer these basic organisational questions effectively. The participants do not have to buy into altruistic ethics or hippie ideals of community. We-Think works; it delivers the goods – whether a new encyclopaedia, a software program, a news service or even different ways of betting against other people. It does this when various vital ingredients are brought together. When they are lacking it will not take off and no one should waste their time trying. A couple more examples will help explain how it works.

The worm *C. elegans* is a simple organism: it has a front end, where the food comes in, a rear where the waste exits, a bottom and a top, a left and right. That is pretty much it, except that even the simplest worm achieves a mind-bogglingly complex task: it generates itself from a set of genetic instructions. The puzzle of how the worm achieves this task was unravelled by a collaborative research effort, which in turn provided the basis for the global, public effort to map the human genome three decades later. Our map of the genome is the product of an elaborate shared authorship. Scientific collaborations like the one behind the unravelling of the *C. elegans* genome are a powerful working model for We-Think culture which the web is helping to spread.

When Sydney Brenner set out to unravel the worm's genome in 1965 – just eight years after Francis Crick and James Watson had uncovered the double-helix structure of DNA – little was known about how genes work. Brenner set out to find how the worm's genes directed the organism's growth, with only a small team of novice researchers and some crude tools: the scientists lifted worms into petri dishes with sharpened toothpicks. It was as if someone had seen the Wright brothers' first flight and decided to start work on a jumbo jet.[1]

Brenner's Laboratory of Molecular Biology provided the community's core. He had the resources to get started and just enough momentum to attract other laboratories to collaborate on a project that intrigued others because it was so ambitious. The working practices of Brenner's small lab set the tone for what eventually became a global project involving thousands of researchers. Brenner's laboratory was hard-working and meritocratic, egalitarian and conversational. People often discussed ideas in the coffee room. They were exploring new territory, devising the process as they went along, so there were no fiefdoms to defend. Sharing ideas quickly became normal. As the community grew, researchers communicated their progress through the relentlessly practical *Worm Breeder's Gazette*. (The *Gazette* was like a cross between *Lean's Engine Reporter*, which organised innovation in the Cornish tin mines and Stewart Brand's *Whole Earth Catalog*, which listed useful technologies.) Brenner's openness set off a virtuous cycle of knowledge-sharing, which was the only way to get the work done. He had identified a task so complex that no single laboratory could complete it. Knowledge about what a particular gene did was worthless unless it could be combined with information about other genes. The jigsaw puzzle had so many pieces it could be completed only through collaboration on a massive scale. Bob Waterston, one of the US leaders of the project, explained:

> The more we put out there the less of a problem it was to get other people to contribute. The more we restricted the flow of knowledge, the more people felt they had to bargain with us before they would release their results. If you just put the data out there then everyone was on the same footing and they were all free to talk about it.

The community grew along with the common store of knowledge it created. In 1975, 10 years after Brenner launched the project, the first international meeting of worm-genome researchers attracted 24 participants. A decade later there was enough information to fill a sizeable textbook. When the complete gene sequence was announced in 1998, US vice-president Al Gore greeted it as the equivalent of the moon landings. By 2002 the worm researchers' meeting attracted 1,600 participants. One thesis listed all 5,000 connections between neurons in the worm's brain. The worm was the most completely understood organism on earth.

Technology played its part in the project's success. By the end, researchers who had started out using toothpicks were working with automated gene-sequencing machines that could do in minutes tasks that in the mid-1960s had taken months. Yet the worm project – like the human genome project that followed in its footsteps – was as much a triumph of social organisation as of technology. Eric Raymond, the guru of the open-source software movement, famously described mass collaborative innovation as a bazaar – open, cacophonous, with no one in control – rather than like cathedral-building, where craftsmen implement a master plan.[2] The worm project could never have been like the building of a cathedral. There was no master plan because no one knew for sure what they would find next or how anything would fit together. Researchers were obliged to fan out and explore, to share their ideas to piece the map together. Yet neither was the worm project like a bazaar. It got going and sustained itself only thanks to Brenner's leadership and a core of contributors. His ambition animated the community and led by example, and his laboratory set the egalitarian, open yet challenging style of working that characterised the whole project. He set the norms, releasing information early to encourage others to do likewise. Brenner made sure all the pieces fitted together to create something of lasting value.

The worm project is a recipe for We-Think. Brenner found

a way to mobilise a vast community of researchers and to combine their different skills and interests with little hierarchy and bureaucracy. His laboratory provided the community's core; the *Worm Breeder's Gazette* and frequent meetings provided a way for researchers to connect and combine their ideas; the open sharing of information allowed thousands of people to collaborate. More popular, non-scientific efforts at We-Think are now replicating this recipe using the web. The most famous example is the open-source software community Linux.

A single grain added to a pile of sand can cause it to collapse. Spotting which grain of sand that will be is virtually impossible. That is perhaps why most of the computer world did not notice in September 1991 when Linus Torvalds, then a wispy computer science student in Helsinki, released onto the Internet the first version of a computer program he had written: Linux. Torvalds did not put just the program online but also its source code – its basic recipe – leaving it for software enthusiasts to take away and tamper with, to criticise and propose improvements. Open-source is software that nobody owns, everyone can use and anyone can improve, and open-source licensing is a way to hold ideas and information in common that under the right conditions can encourage mass collaborative innovation.[3] That is what Torvalds eventually set off – an experiment in geek democracy, people thinking and creating together, which has created a program that is complex, robust and now widely used. The Linux community is also the most impressive example of sustained We-Think, with ideas shared among a large community over more than 15 years to develop a highly sophisticated but reliable product. Linux shows that sharing makes sense, especially when ideas are at stake and the web is in play.[4]

Like many radical innovations, Linux is not as new as it

sounds.[5] It draws on more than four decades of innovation in sharing software and computers, but it arrived at just the right time. Linux drives Internet applications, and as the Internet spread, so demand for it grew. In the five years from 1999 the number of Linux installations grew by 28 per cent a year, and by 2006 Linux accounted for about 80 per cent of software on computer servers worldwide.[6] The figure of 29 million registered Linux users in the world by 2006 vastly underestimates the true user base for open-source software.[7] Every time anyone runs a search on Google they are a Linux user because Google's servers run on Linux.

Linux's market share expanded as the software became more complex. The version distributed in March 2000 by Red Hat, which specialises in installing Linux for corporations, had 17 million lines in its source code. One estimate suggests it would have taken 4,500 person-years of work for professional software-coders to develop this, at a cost of about $600 million.[8] The version released by Debian in June 2005 had 229 million lines in its source code and would have taken 60,000 person-years of software-coding at a cost of perhaps $8 billion.[9] (To put that in context, the version of Microsoft Windows XP released in 2002 had an estimated 40 million lines in its source code.) IBM is just one software company that has recently started investing in shared, open-source platforms such as Linux. IBM executives estimate it would have cost the company ten times as much itself to develop Linux to the stage at which it is now as it has to participate in the community that developed it.

The Linux community has sustained this phenomenal growth by adopting the organisational features of We-Think. At the core is a small band of trusted programmers working closely with Torvalds, a quietly inspirational leader guiding

CORE
CONTRIBUTE
CONNECT
COLLABORATE
CREATE

by example. Membership of this core group, which looks after the program's kernel and its future development, is earned by people who put in long hours of high-quality volunteer programming. In 1994 there were just 80 people in the core; by 2001 there were 400. A much larger community of users and contributors has formed beyond this. By 2007 there were 655 Linux user groups in 91 countries, sharing ideas through websites and bulletin boards as well as face to face at conferences.

Were Linux a one-off it would be interesting and exotic. But it is not. Most of the Internet relies on open-source software, much of it created through collaboration: most websites rely on servers running the open-source Apache program; MySQL is an open-source database program; Perl and Python are open-programming languages. There are many more in their wake, with names such as Dupral, Evolt, Tomcat and JBoss. In 2007 the directory Sourceforge.net listed more than 90,000 open-source initiatives.[10]

Linux has succeeded as a product only because the community that supports it has organised itself systematically to create, share, test, reject and develop ideas in a way that flouts conventional wisdom. Successful We-Think projects are based on five key principles that were all present in both Linux and the worm project.

Core

Everything has to start somewhere. Somebody has to be willing to work harder than everyone else or nothing ends up getting done. Innovative communities invariably start with a gift of knowledge provided by someone, just as Linux started with the kernel that Linus Torvalds slaved over and which he posted on the Internet.

A good core attracts a community of capable contributors and developers around it. The kernel has to be solid but unfinished, so open to improvement; if it were already complete there would be few opportunities to add to it. Jane McGonigal says the core to a successful game like I Love Bees depends on the starting-point being ambiguous and open to interpretation. Both the worm project and I Love Bees began with a puzzle that could be solved only with the collaborative efforts of people with different skills. Steven Weber, a political scientist at Berkeley University in California, found that successful open-source software projects tended to be 'multi-dimensional' and complex, thus inviting the involvement of people with different skills.' Thomas Kuhn summed up the ambiguous character of the core to a new intellectual community in his history of scientific revolutions. Kuhn argued that the possibility of a new scientific paradigm emerged when a small group of pioneers made a breakthrough that was

> sufficiently unprecedented to attract an enduring group of adherents away from competing modes of scientific activity. Simultaneously, it was sufficiently open-ended to leave all sorts of problems for the redefined group of practitioners to resolve.[11]

However, a core will develop only if its creators give away the material on which others can work, to which they can add and which they can refine. Successful innovation comes from a creative conversation between people who combine their different skills, insights and knowledge to explore a problem.[12] We-Think is creating a new way for these conversations to emerge. A good core starts a creative conversation, and invites people to contribute.

Contribute

A successful creative community has to attract the right mix of people, who have different ideas and outlooks and access to tools that enable them to contribute. We-Think takes off only by getting the right answer to each of the following questions: Who contributes? What do they contribute? Why do they do so? And how do they do it?

Creative communities have a social structure. As we have seen, a relatively small, committed core group tends to do most of the heavy lifting: the discussion moderators in Slashdot; the original inhabitants of Second Life. These are the Web 2.0 aristocracy: people who because they have been around longer and done more work tend to get listened to more. There is nothing unusual in this. Most innovative projects, whether inside a company or a theatre group or a laboratory, start with intense collaboration among a small group which shares a particular passion or wants to address a common problem – as did the worm researchers who gathered around Sydney Brenner at Cambridge.[13] Often, however, such communities can become closed and inward-looking. To be dynamic, they have to open out to a wider world of more diverse contributors who add their knowledge or challenge conventional wisdom.

We-Think projects take off when they attract a much larger crowd, who are less intensely engaged with the project. Their occasional, smaller contributions may in aggregate be as significant as the work initially done by the core. Linux, for example, as well as having 400 key programmers at the core, has close on 150,000 registered users – akin to members – who may only report the occasional bug in the program. Yet such a report may provide the starting-point for a much more significant effort at innovation. The make-up of the crowd is as important as the brainpower of the highly committed core. Crowds are

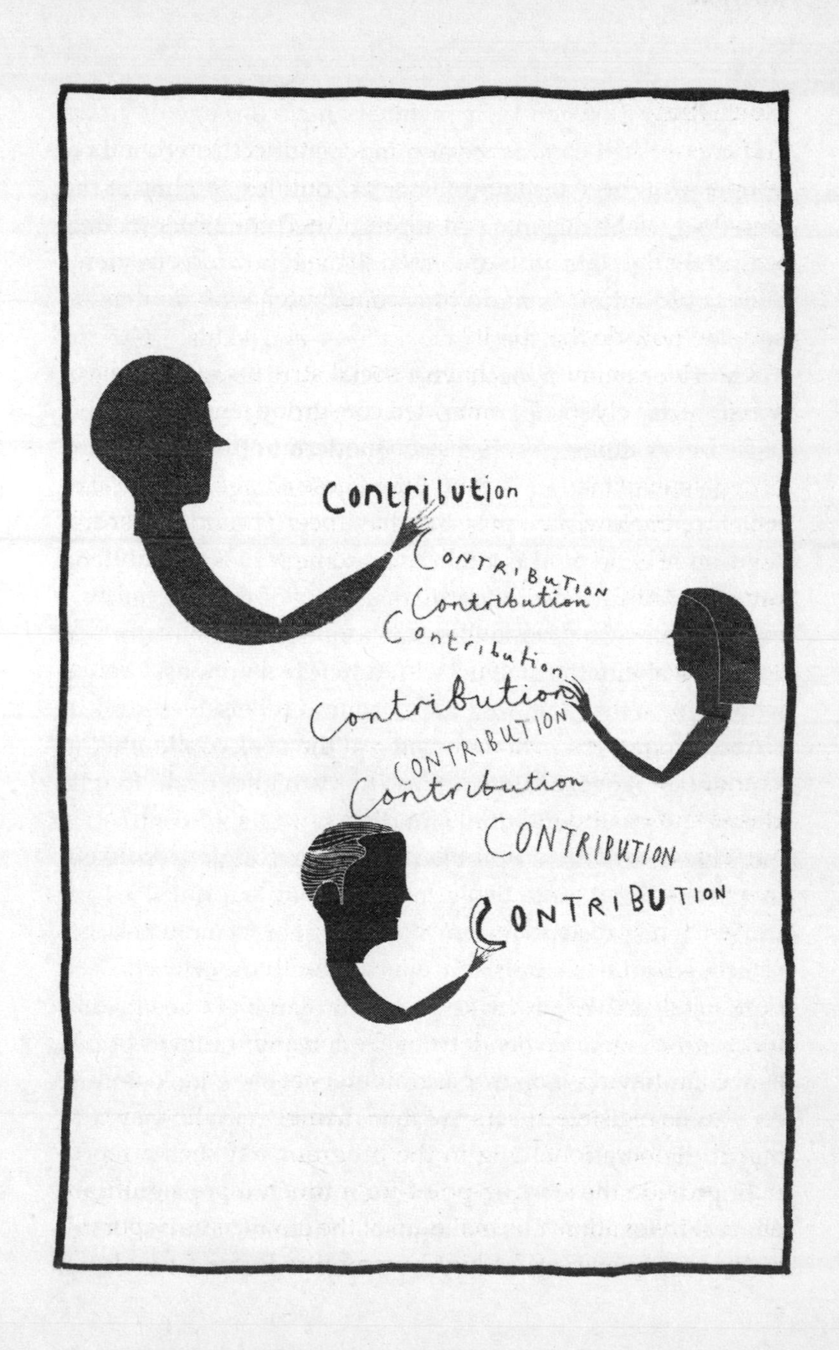

Contribution
CONTRIBUTION
Contribution
Contribution
Contribution
CONTRIBUTION
Contribution
CONTRIBUTION
CONTRIBUTION

intelligent only when their members have a range of views and enough self-confidence and independence to voice their opinions.[14] Scott Page, a professor of complex systems at the University of Michigan, used sophisticated computer models to find that groups with diverse skills and outlooks came up with smart solutions more often than groups of very clever people who shared the same outlook and skills.[15] Groups made up of many people who think in different ways can trump groups of people who are very bright but very alike, Page argues, so long as they are organised in the right way.

Page's explanation is that the more vantage points from which a complex problem is seen, the easier it becomes to solve. A group of experts who think in the same way is probably no better at devising a solution than just one of them, so adding more people who think in the same way is unlikely to improve a group's ability to come up with different solutions. Groups who think in the same way can often find themselves stuck at the same point – akin to their being at the peak of a foothill in a mountain range, unable to climb to the higher peaks that lie beyond. A group who thinks in diverse ways, in contrast, is more likely to address a problem from many angles, less likely to get stuck and more likely to find a way out if it does get stuck. Diverse viewpoints are likely to generate more possible solutions and evaluate them in a wider range of ways. The right perspective can make a difficult problem seem easy. Innovation often involves trying out many vantage points before finding the one that makes the problem look simple. As Thomas Edison put it, 'We have found 1,000 ways not to make a light bulb.'

Bugs in a software program often become apparent only when the program is tested in many different settings. Better 1,000 people making different tests at the same time than

a single person making 1,000 tests one after the other. This explains why open-source programs are often more robust that proprietary software: they have been tested much earlier by a much wider group of users. Bart Nooteboom, a professor at Rotterdam University, argues that distributed testing of this kind is vital to most innovation. He examined the development of 17th-century Dutch sailing ships and found that the designs mutated when the community of sailors tested and then adapted them to meet different conditions: first canals, then lakes, larger inland waterways, offshore sailing, the North Sea, the Atlantic and so on.[16] We-Think allows ideas to be tested from a larger, more diverse set of vantage points more quickly and with ideas continually passing between the tightly knit core who develop them and the crowd who test them out.

This testing becomes possible only when people can make the kind of contributions they feel happy with, which requires tools to allow them to get involved. Mass computer games thrive by making it easy for player-developers to pick up tools to create content. Blogging depends on easy-to-use software for writing and publishing online. The camera phone is now a ubiquitous tool for citizen journalism. Such tools are taking to mass scale the self-help ethic of the original computer hackers. The first versions of the Unix operating system, on which Linux is based, were created by lone programmers who could not afford to provide tech support to their users. So when they sent their programs to people, usually on a stack of floppy disks, they included a set of tools that allowed users to sort out problems themselves.[17] When people can get hold of tools that allow them to produce aspects of a service, they start becoming players, participants and developers: newspaper readers become writers, publishers and distributors; bystanders become photographers; the audience can become reviewers and critics.

Perhaps the most perplexing question is not *how* people contribute, but *why* they do so – particularly when they are not being paid and their work is given away. In open-source software projects, a few are inspired by a hatred of proprietary software providers, especially Microsoft.[18] A minority are driven by altruistic motives. Some see their involvement as a way to get a better job: by showing off their skills in the open-source community they can enhance their chances of being employed.[19] For the majority the main motivation is recognition: they want the acknowledgement of their peers for doing good work that they enjoy, that gives them a sense of achievement and in the process solves a problem for which other people are seeking a solution. Many of the most striking Web 2.0 success stories started when users created tools to solve a problem they themselves faced – keeping track of all the blogs being created, sharing video and photographs online – and which quickly got taken up by others who faced similar problems.

Open-source gives away intellectual property so other people can freely use it. We-Think requires more than that: it is also an invitation to participate and collaborate in creating something. Open-source ownership of a project becomes powerful when it enables mass collaborative approaches to innovation. For that to be possible many ideas have to be combined; contributors have to meet and connect with one another.

Connect

At the St Louis world fair in 1904, an ice-cream stand ran out of cups. The owner of the waffle stand next door started rolling his waffles to form cones. There was nothing new in either ingredient, but the combination of ice-cream and the waffle created something entirely new. The more combinations a community can create, the more innovation there will be. Cities are creative when they make these combinations possible. The same is true of We-Think.

Diversity counts for little unless the different ideas that are floating around can be brought together to cross-pollinate. A community that is diverse but Balkanised will not be creative. People with different ideas must find a way to connect and communicate with one another. When they do, and in the right way, the results can be explosive. James Watson and Francis Crick unravelled the double-helix structure of DNA because they found a way to combine their very different outlooks. Crick's training spanned physics, biology and chemistry. Watson had trained as a zoologist but had become fascinated by DNA after studying viruses. They combined their ideas through constant, intense conversation of a kind of which their rivals were incapable. Watson and Crick's collaboration was a case of one plus one equals twelve.[20]

The larger the group and the more diverse perspectives are involved, the greater the benefits from combining them. Take five people, each with a different skill. That gives 10 possible pairings of skills. Add a sixth person with a different skill. That creates not 12 pairs but *another* five possible pairings 1 – 115 in all. A group with 20 different tools at its disposal has 190 possible pairs of tools and more than 1,000 combinations of three tools. A group with 13 tools has almost as many tools – 87 per cent – as a group with 15 tools. Not much of a gap. But if a task requires combining four tools it is a different story. The group with 15 tools has 1,365 possible combinations of four tools. The group with 13 tools has 715, or about

52 per cent. Groups with larger sets of diverse tools and skills are at an advantage if they can combine effectively to take on complex tasks.

Markets are not the best way for people with diverse skills to connect and combine. A market might provide a way for someone with a problem to find someone else who might have a solution: if you have a leaking tap, you look for a plumber. That is the model of Innocentive, the scientific problem-solving community that was spun off from the drugs company Eli Lilly. Companies can post their scientific problems on Innocentive's website to see if they can be solved by one of the more than 100,000 scientists signed up to the market. But markets of this kind have inherent limitations: they work for specific problems that need exactly the right individual to solve them. They do not provide the basis for sustained creativity and innovation to explore difficult complex puzzles. That is a kind of problem-solving that comes only from intense collaboration. In the worm project, the researchers started by meeting in the coffee room at Brenner's laboratory. In We-Think, crowds need meeting-places, neutral spaces for creative conversation, moderated to allow the free flow of ideas. This is why, at their heart, these projects have open discussion forums and wikis, bulletin boards and community councils, or simple journals like *Lean's Engine Reporter* and the *Worm Breeder's Gazette*, so that people can come together in a way that allows one plus one to equal twelve many times over.

In We-Think projects, the task of combining ideas is made easier because the products usually fit together like Lego bricks: they are made from many interconnecting modules. Modularity is not new; it has been a feature of computer development since at least as long ago as the 1960s, when IBM was developing its System/360 computer. Fred Brooks, the person responsible, wanted everyone involved to be kept abreast of what everyone else was doing. Daily notes of changes to the program were shared with everyone. Quite soon people were starting work each day by sifting through a two-inch wad of notes on these changes. The costs of communication and co-ordination spiralled out of control. Miscommunication and misunderstandings grew. Adding people to the project did not solve the problem: more work got done, but more misunderstandings were created and with them more bugs. When the wad was five feet thick, Brooks decided to break the S/360 into discrete modules that could be worked on separately. A core team set some design rules specifying what modules were needed and how they should click together. This meant that module-makers could concentrate on their patch while the core team looked after the architecture of the system as a whole. New and better modules could be fitted into the system without its having to be redesigned from scratch.[21]

Modularity really pays dividends when it is combined with open ways of working – when it enables a mass of experiments to proceed in parallel, with different teams working on the same modules, each proposing different solutions. This combination is how open-source gets the Holy Grail: a mass of decentralised innovation that all fits together. Just as Lego bricks come in a dizzying array of colours, shapes and even sizes but all have the same system of connectors, We-Think projects have rules for making connections that usually come from the core team. This is what allows a mass of independent but interconnected innovation. Mass computer games, collaborative blogs, open-source programmes and the human genome project all share this feature: they click together masses of modules.

However, a Lego brick structure is not enough to make

We-Think work. Groups also need to make decisions. Diverse contributors can combine their ideas only if they can agree how to collaborate. Any commons will fall into disrepair if it is not effectively self-regulated. That is far easier said than done.

Collaborate

A mass of contributions does not amount to anything unless together they create something ordered and complex. An encyclopaedia is not a mass of random individual contributions; it is a structured account of knowledge. People playing a game or building a community need to agree rules to govern themselves, or chaos ensues. How do We-Think communities govern themselves without an obvious hierarchy being in charge, enforcing the law? This challenge is not technical but political. We-Think works only when it has responsible self-governance, and that is a particularly difficult thing to achieve in highly diverse communities.

People often think in different ways because they have very different values; what matters to them differs. Someone who sees the world through art and images will acquire skills – drawing and painting – that make it easier for them to work. Someone who sees the world in terms of numbers and money is more likely to become an accountant, to use a calculator rather than a paintbrush. A large toolbox that includes both calculators and paintbrushes, both artists and accountants, is good for innovation.

The trouble is that people with fundamentally different values often find it difficult to agree on what they should do and why. Diverse ways of thinking are essential for innovation; diverse values, based on differences about what matters to us, often lead to squabbles. This is why diverse communities often find it more difficult agree on how to provide public goods, such as healthcare, welfare benefits and social housing. Diverse groups can become very unproductive when their differences overwhelm them, provoking conflicts over resources or goals. Elinor Ostrom found that shared fisheries, forests and irrigation systems required effective self-governance and local monitoring by participants to make sure no one was over-using resources. When local self-governance fails, the commons collapses and innovation becomes impossible.

We-Think succeeds by creating self-governing communities who make the most of their diverse knowledge without being overwhelmed by their differences. That is possible only if these communities are joined around a simple animating goal, if they develop legitimate ways to review and sort ideas and if they have the right kind of leadership. What they are not, ever, is egalitarian self-governing democracies.

As an example, consider the open-source community that produces Ubuntu, a user-friendly version of Linux. Mark Shuttleworth, Ubuntu's founder, is like a benevolent dictator and reserves some decisions for himself, such as the design of the Ubuntu website. The heart of the community, the technical board, meets online to set technical standards and to define what should be included in the different versions of the program. The board's decision-making is transparent and open: anyone can propose additions to policies through the Ubuntu wiki; the board's agenda is made available as a wiki every two weeks; and anyone can attend the online meetings as an observer. The decisions are taken, however, by Shuttleworth and four other board members, whom he appoints – albeit subject to a vote among the community's lead programmers. Meanwhile a separate Ubuntu community council supervises the social structure, creating new projects and appointing leaders for teams that support different releases and features

Share your ....
No?
Did you HEAR about....?
....Thoughts

of the program, such as those for laptop users. Then there are the LoCo teams around the world who promote the use of Ubuntu in their country. Someone can become an Ubuntu member (an Ubuntero) by coding software, documenting changes, contributing artwork or acting as an advocate for Ubuntu. In mid-2007 the community had 283 core members. Those with most power and responsibility – dubbed Masters of the Universe – are the core developers and they have their own council to determine who should be allowed into their guild.

The lesson of Ubuntu – which is still far from a proven success – is that effective governance of creative communities is like a lattice-work. Decision-making is very open: anyone can see what is decided and how; anyone can make suggestions about what should be done. But the way decisions are made is rarely democratic. Ubuntu the product may be open-source; the community that sustains it is far from open-ended. These are not like the Utopian communes of the 1960s – which is why they might be more successful than co-operatives of the past.

Create

We-Think enables a mass social creativity which thrives when many players, with differing points of view and skills, the capacity to think independently and tools to contribute, are brought together in a common cause. If the players are distributed they must have a way to share, combine and cohere around a common goal. However, for much of the time contributors may work independently and in parallel, often reworking elements of a core central product – whether that's an epic poem in Ancient Greece, a piece of genetic code, a latter-day software program or an encyclopaedia. The product grows through accretion and a reciprocal process of observation, criticism, support and imitation. Most people take part because they get an intrinsic pleasure from the activity and seek recognition from their peers for the work they have done. These communities must have places – forums, websites, festivals, gazettes, magazines – where people can publish and share ideas. Social creativity is not a free-for-all; it is highly structured. Although the lines between expert and amateur, audience and performer, user and producer may be blurred, those with more standing in the community, based on the history and quality of their contribution, form something like a tightly networked craft aristocracy. Social creativity collapses without effective self-governance: decisions have to be made about what should be included in the source code, published on the site, pushed to the top of the news list. Participants who do not abide by the community's rules have to be excluded somehow. They must respect the judgements of their peers.

The raw material of these collaborations is creative talent, which is highly variable. People are good at different things and in different ways. It is difficult to tell from the outside, for example by time-and-motion studies, who is the more effective creative worker. It is impossible to write a detailed job description for a creative position specifying what new ideas need to be created by whom and by when. Open-source communities resolve the difficulties of managing creative work by decentralising decision-making down to small groups who decide what to work on, depending on what needs to be done and the nature of their skills. It is very difficult for someone to pull the wool over the eyes of their peers; they will soon be found out. When it works, peer review excels at sharing ideas and maintaining quality at low cost.

Conclusions

We-Think will not work where there is no core around which a community can form; where experimentation is costly and time-consuming, and so feedback slow; where decision-making becomes cumbersome or opaque, beset by complex rules; where the project fails to attract a large and diverse enough community. It will not take off if tools to add content are difficult to use; if contributors cannot connect to one another; if communities cannot govern themselves effectively and so either fracture or ossify. For many important activities, We-Think will make no sense at all: performing medical operations, cooking meals, running nuclear reactors, railways or steel mills. It is not well suited to tasks where only professional expertise will do. In late 2006 I had a minor operation and was very glad to find that the surgeon was not assisted by a group of pro-am butchers, bakers and candlestick-makers who were taking their lead from the Wikipedia entry on the procedure they were about to perform. (Pro-ams are people who undertake activities as amateurs but to professional standards.)

We-Think works only under certain conditions. Usually, a small group creates a kernel which invites further contributions. Its project must be regarded as exciting, intriguing and challenging by enough people with the time, means and motivation to contribute. Tools should be distributed, experimentation cheap and feedback fast, enabling a constant process of trialling, testing and refinement. The product should benefit from extensive peer review, to correct errors and ratify good ideas. Tasks should be broken down into modules around which small, close-knit teams can form, allowing a range of experiments to run in parallel. There should be clear rules for fitting the modules together and separating good ideas from

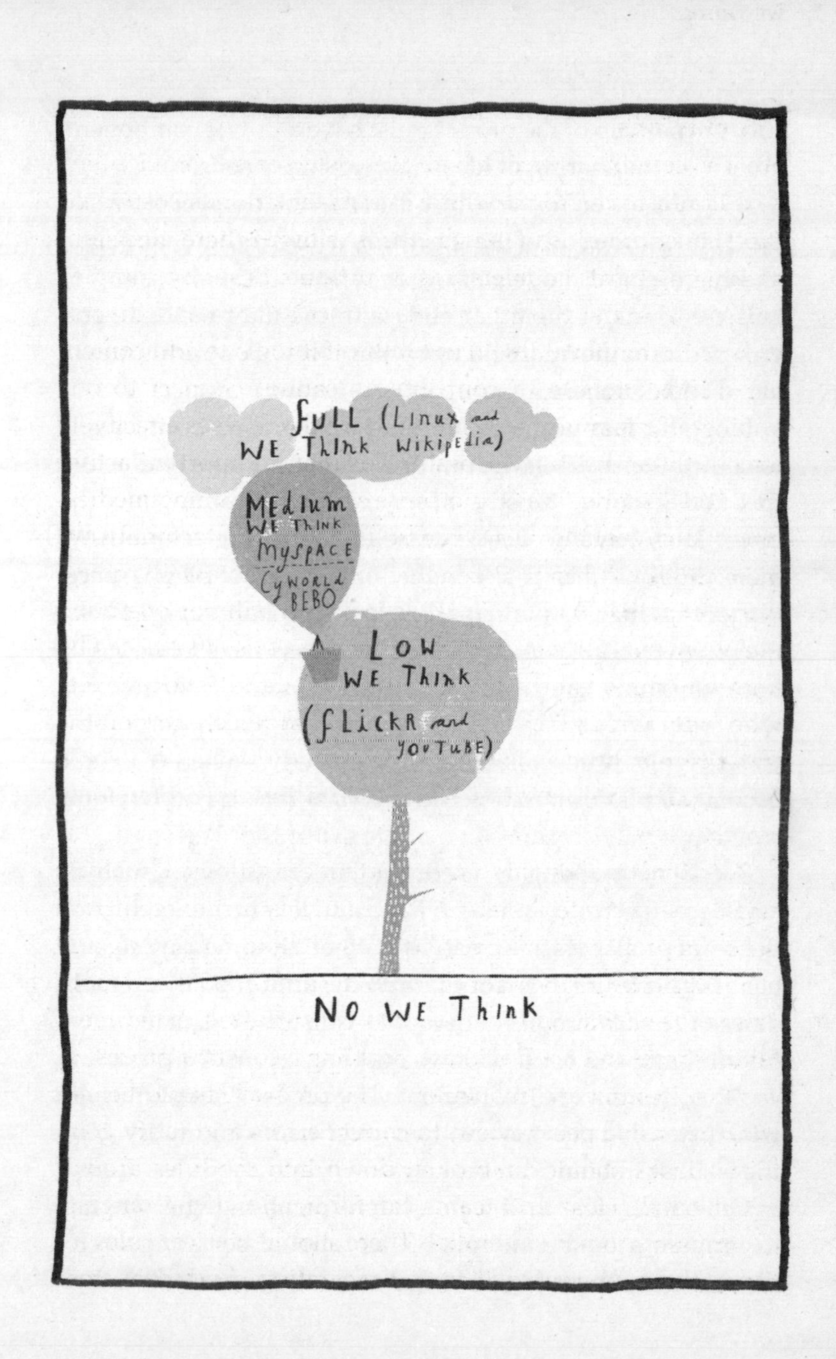
FULL (Linux and
WE Think Wikipedia)
MEDIUM
WE THINK
MySPACE
(Cyworld
BEBO
LOW
WE Think
(FLICKR and
YOUTUBE)
NO WE Think

bad. Ownership of the project must have a public component, otherwise the sharing of ideas will not make sense.

It is not all or nothing but a matter of degree: from No We-Think at one end of the spectrum, where traditional, closed and hierarchical models of organisation still make sense, to Full We-Think at the other end, with the likes of Linux and Wikipedia. In the middle, there will be lots of opportunities to blend some of these ingredients in different ways.

Blogging is a prime example: it allows a mass of people to contribute their views, but only rarely do they find a core to build around. Mostly, bloggers communicate into the ether. They have no desire to build something with others, merely to leave their mark on their little patch of digital space. Blogging is high on participation, low on collaboration. Flickr, the photo-sharing site, and YouTube, the video site, fit in this Low We-Think category: they allow a mass of participants to connect with an audience and with one another. Yet there is relatively little collaborative creativity. When YouTube becomes a platform for people to collaborate in making films together it may acquire some of the features of We-Think.

Social networking is Medium We-Think. Sites such as MySpace, CyWorld and Bebo have not yet encouraged much deliberate collaborative creativity, although some participants have begun to use them for example to support political candidates or to rally around causes they care about. Collaborative filtering and the book reviews and ratings on Amazon, and social tagging tools like Technorati and del.i.cious, through which people help one another find interesting material on the web, fit into this category.

Only when all our five conditions come together at scale to provide a deliberate, conscious form of social creativity in which many people contribute and collaborate does Full

We-Think emerge. OhmyNews, the South Korean citizen-journalist news service fits in here, as do mass computer games like World of Warcraft and scientific collaborations like the project to unravel the worm's genome. Full We-Think is the deliberate and organised combination of contributions from a mass of distributed and independent participants.

It would be silly to suggest that We-Think can work in every situation and that it is always the best organisational recipe. The challenge is to engage in more We-Think when it is appropriate, which is when we are collectively trying to solve a complex problem, or to create something that no individual could produce and where creative thinking is critical to develop ideas. We-Think will not touch all organisations but some will be transformed, and many will find some aspects of what they do changed, possibly quite fundamentally, by this new organisational recipe. How We-Think will change the way we run, lead and own organisations is the subject of the next chapter.

4
THE WE-THINK BUSINESS

Industrial-era corporations are at war with themselves and we are caught in the crossfire. The endless restructuring and downsizing of the last three decades provides the backdrop for the rise of We-Think as an alternative way to imagine how we could organise ourselves.

The large corporation that emerged in the late 19th and early 20th centuries was built on a military model of organisation: everyone had their place in a rank, every place defined a function, and authority flowed through a chain of command from top to bottom. If you were unsure what to do next, you looked at your job description and if that did not provide an answer you asked the next person up in the chain of command for instructions. Industrial-era organisations were in Max Weber's phrase 'iron cages', forcing people to conform to the power of bureaucracy. Since the late 1970s the bars have been whittled down, bent and removed. As corporations struggled to accelerate innovation, improve quality, cut costs and attract consumers with a wider array of products, it was often not clear whether the hierarchical corporation was disintegrating into networks of outsourced production, or whether on the contrary it was tightening its grip through stricter performance management and more centralised control of brands. The outcome is that in most corporations hierarchies are now flatter, job descriptions vaguer, the working day more flexible, the working week more extendable, careers more unpredictable and the boundaries of organisations more porous as companies have to come to rely on shifting networks of polygamous partners and suppliers. Getting more productivity, innovation and quality seems to require ever greater pain to make organisations leaner and meaner. The outcome is more managerial turnover, employee stress and organisational turmoil.

We-Think is emerging as an alternative organisational recipe because it provides a more effective answer to the multiple challenges organisations face. We-Think offers a different approach to innovation by encouraging the free sharing of ideas that come from multiple sources within and beyond the company; to work, by deploying self-management to make it more efficient and creative; to consumption, by turning consumers into participants in creating solutions and so mobilising their commitment, effort and ideas; and to leadership, which must mobilise communities rather than concentrating power at the top and issuing instructions from on high. Open and collaborative models of organisation will increasingly trump closed and hierarchical models as a way to promote innovation, organise work and engage consumers.

Corporate efficiency, especially in America, has been bought at the cost of a growing sense of social dislocation.[1] People trust corporations less. Careers have become more fragmented. A job is now a set of tasks rather than a craft demanding devotion. Relationships have been turned into transactions. Leadership has become little more than bonus-driven performance management. Experience is prematurely discounted in a world where novelty is everything. Impatience is the hallmark of corporate life. Workers and parents, consumers and managers are living like circus performers, juggling and jumping their way through life. The communities that form around We-Think projects seem so appealing because they offer to renew the fractured social contract underpinning work and production. The impatient, flexible corporation carelessly writes off people, relationships, experience and community. We-Think offers a way for capitalism to recover a social – even a communal – dimension that people are yearning for.

In the last decades of the 20th century the market and corporations were triumphant and co-operative and collaborative values were in retreat. We-Think is emerging as a reaction to this: in field after field large groups of committed and knowledgeable contributors, often amateurs, many working without pay and collaborating with little hierarchy, are mobilising resources on a scale to match the biggest corporations in the world to create complex products and services. It should not be possible. Pigs do not fly. In the next few years the irresistible force of collaborative mass innovation will increasingly be meeting the immovable object of entrenched corporate organisation.[2] This chapter is about the kind of organisations that could emerge from that collision.

The conflict will not be one-sided. We-Think certainly will not sweep the board and transform every business. The struggle between organisational models will expose the shortcomings of some of the early models of We-Think that have inspired such optimism. The wiki-economy has not escaped the deep-seated problems that beset earlier attempts at collaborative endeavour. Communes, mutual societies and worker co-operatives often failed because they closed in on themselves and avoided hard decisions about how work should be organised and money made. We-Think has produced some impressive collective voluntary initiatives, but most people cannot get their groceries and children's clothes from the gift economy. We-Think will not spread far nor sustain itself if it is confined to tasks for which people are prepared to volunteer. People must find a way to make their livings from these collaboratives and invest in them. We-Think entrepreneurs are consequently desperately searching for viable business models that will allow them to earn some money without turning their backs on community values, while traditional companies are searching for ways to become more open and collaborative. The We-Think gift economy needs to find an accommodation with the market economy in which goods and services have to be paid for. The most exciting business models of the future will be hybrids that blend elements of the company and the community, of commerce and collaboration: open in some respects, closed in others; giving some content away and charging for some services; serving people as consumers and encouraging them, when it is relevant, to become participants.

We-Think will gradually change five fundamental aspects of economic life: how we work, consume, innovate, lead and own productive endeavours. Let's start with innovation.

Innovation

Modern capitalism is defined by its ability to conjure a stream of new products, services, organisational models and experiences almost out of thin air. For most of the 20th century we believed we knew that new ideas would come from special people, working in special places, often wearing special clothes: the boffin in his white coat in the lab; the artist in his smock in the studio; the zany inventor in his garage; the loft-living bohemian wandering the cultural quarter. If you wanted more creativity in your company you needed more people and places like this: large R & D labs or 'skunk' works in forests and woods, where researchers laboured behind barbed-wire fences and security gates. Large companies needed pipelines to link these centres of creativity to manufacturing and markets. The pipeline model rests on the idea that every innovation has a moment of birth, usually when the idea springs into the head of its inventor, who is uniquely positioned to say what his invention is for and how it should be used. That purpose is then enshrined in a patent. Innovation was seen as a linear and sequential process from invention through development to application. The consumers waiting at the end of the pipeline played little role other than deciding whether or not to use the product.

The pipeline model of invention-driven innovation is increasingly unproductive.[3] In the pharmaceutical industry, where R & D is most intensive, the costs of developing and testing a blockbuster drug have risen to close to $1 billion. Applications to the US Food and Drug Administration, the agency that evaluates new pharmaceuticals, fell from a high of 131 in 1996 to 72 in 2003. In 2002, the FDA approved just 17 new molecular entities, and in the following year 21, of which only nine were judged to be significant. No major new antibiotics have been approved for 40 years. Many long-term conditions, including cancer, lack treatments. Diseases of the poor are largely neglected. Yet the drug research pipeline is not drying up for want of resources. Between 1995 and 2002, US pharmaceutical companies' R & D nearly doubled, to $32 billion. Drug development costs are rising faster than inflation. Between 1980 and 2005 pharmaceutical-industry revenues rose by about 11 per cent a year, but development costs rose by close to 15 per cent a year. Meanwhile, the window of opportunity for companies to exploit their inventions by charging a premium price is getting narrower. In the 1970s, a drug company might have had 20 years to earn back its investment in a patented drug that had taken 10 years to develop. Now, a patented drug can face competition within 10 years from an alternative or a near copy.

Rising development costs, less certain returns and more competitive markets mean innovation is an increasingly fraught business. It should be no surprise that pharmaceutical companies and semi-conductor companies are leading the way in exploring how to cut research costs through collaboration. As the closed, pipeline model of innovation has come under growing strain, so We-Think has emerged to provide an alternative model of innovation as an open, even a mass, activity. Open innovation uses the web to take to scale a collaborative and social approach to creativity.

Most creativity emerges when different points of view are held in reciprocal tension, so that they play off one another, eventually evolving into a new idea. The idea for Henry Ford's revolutionary moving assembly line did not simply spring to life in his head in a darkened room after months of inner reflection. Ford's innovation came from a team who borrowed and blended ideas and techniques: from a machine-tool industry that used interchangeable parts; from meat-packing which used a moving line to cut up carcasses; and – for scheduling techniques – from the railroads. Most advances in 20th-century science came from creative conversations that blended ideas. Werner Heisenberg's conversations with Neils Bohr and other physicists in Copenhagen in the 1920s paved the way for quantum mechanics and other theories that led not just to the nuclear bomb but to many advances in modern electronics. Even Thomas Edison, the most famous lone inventor, owed his success to his being a great collaborator, a skill he

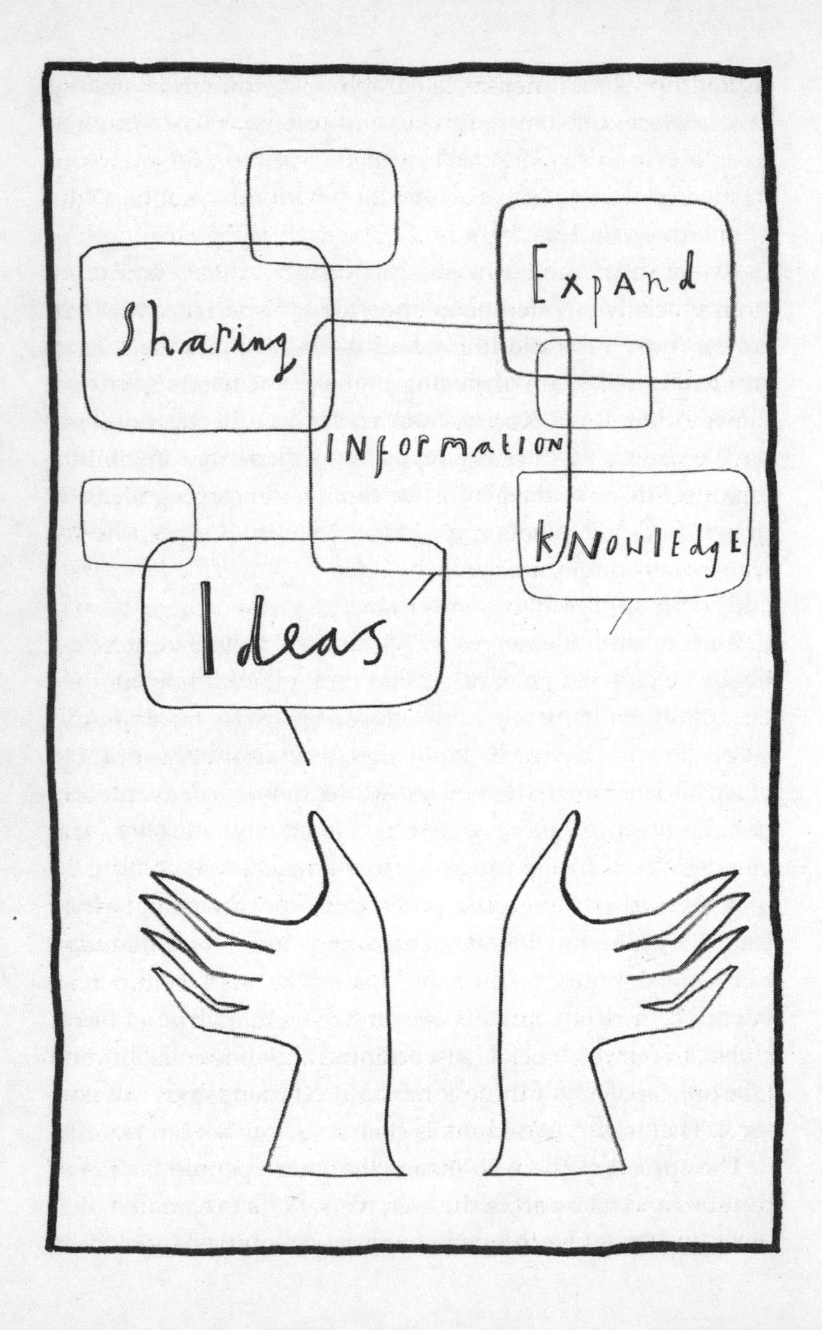
Sharing
Expand
Information
Knowledge
Ideas

picked up as an itinerant telegraph operator, rarely staying in one place, constantly mixing and mingling with different people. Edison's laboratory in Menlo Park, New Jersey, which opened in 1876 to be a 'factory for invention', produced the phonograph and the light bulb. The mythology surrounding the Menlo Park lab enshrined the idea that innovation came from specially talented people working in special conditions, cut off from the rest of the world. Yet Edison acknowledged that without his team of unsung engineers – Charles Batchelor, James Adam, John Kuresi, Charles Wurth – he would never have come up with many of the inventions that made him famous. Edison had a genius for rapidly developing ideas by drawing on the talents and skills of others. His laboratories were small communities of creativity.

Innovative combinations emerge from intense and extended creative conversations in which ideas are opened up to challenge but also built upon. London coffee houses were vital public forums for such conversations in the 17th and 18th centuries. Edward Lloyd's coffee shop, which encouraged a focus on trade, shipping and insurance, eventually became the insurance market of the same name. The New York Stock Exchange had similar origins. As Mark Ellis puts it in *The Coffee House: A Social History*, the convivial, bawdy coffee house was where 'men learnt new ways of combinatorial friendship turning their discussions there into commercial ventures, critical tribunals, scientific seminars and political clubs'. In our day, social networking and online collaboration take our capacity for these combinatorial friendships to a new level. That is why We-Think is changing how we innovate.

The spread of the web means that more people than ever can start and take part in these conversations to combine their ideas and insights. In open-source communities, innovation succeeds through early exposure to comments and criticism, which allow ideas to be refined, adjusted and reinterpreted. Open-source communities provide a setting for critical, creative, often raucous and sometimes brutally honest discussion. Creativity will still emerge from specially gifted individuals working in special places. But thanks to rising educational attainments, spreading communications and cheaper technology, innovation and creativity are becoming increasingly distributed and emerging from many, often unexpected, sources. Henry Ford created a model for mass production; Linus Torvalds and his ilk are creating a way to organise mass innovation.

Economic conditions increasingly favour these open innovation models, with ideas coming from more sources. In the 20th century most innovation came from large manufacturing companies with research and development facilities. But in the US, for example, the share of R & D done by small companies rose from 4.4 per cent in 1981 to more than 25 per cent 20 years later. Spending on R & D is rising fast in China and India. Skilled labour is increasingly mobile and available all around the world. Places that used to be on the technological periphery, such as India, are closely connected to the core of the world economy. Many innovations now combine multiple technologies and the know-how of different disciplines. All of this is encouraging large companies to shift towards more collaborative, networked approaches to innovation to share costs and multiply their sources of ideas.

The consumer-goods company Procter and Gamble (P & G), for example, has set itself the goal of getting 50 per cent of its new ideas from external sources. When it embarked on its open-innovation programme, P & G estimated it had about 7,500 in-house researchers, compared with the 1.5 million researchers worldwide with skills relevant to its business. By 2006, the company still had 7,500 internal researchers but also drew on 2,000 researchers among its suppliers and 7,000 virtual or extended relationships. P & G is also letting go of more of its ideas to see what other companies can do with them. All its patents will be released five years after they are lodged or three years after a product is shipped. IBM has been through a similar transformation, donating more than 500 software patents to the Open Source Foundation and funding Linux development with at least $100 million a year. Nokia, the mobile-phone company, has attracted more than 1 million developers to its online developers' forum and announced it will not take action against open-source applications of its patents. Lego has started to work closely with software-developers creating open-source programs for its Mindstorms suite of products. Google, Yahoo and Intel have all developed ways to work with large developer communities.

More companies will create open-innovation models that draw on the ideas of communities outside the company and share chunks of intellectual property that was once jealously guarded. These corporate models of We-Think will run alongside more self-governing, hacker communities such as Wikipedia and Linux, which are developing business models to sustain themselves. Scientists, educators, researchers in any field now have the option of trying to create a We-Think-style approach to sharing and developing ideas. At one end of the spectrum there will still be organisations that rely on the pipeline model with new ideas still coming from specially talented people working in special places. At the other end there will be more attempts to create open, We-Think communities. The most fertile space will be in the middle, where many organisations, public and private, will be trying to find the most productive way to share ideas while also making money from them, mixing commerce and community. More companies will attempt to follow Procter and Gamble to create an innovation process that casts a wide net to draw more ideas into their organisation from outside. Even more intriguing, however, will be efforts like Linux which are attempts to use a common core to create an entirely new community. The first approach is open innovation by the sucking of ideas into a company; the second is open innovation by the giving away of ideas that then spread and multiply.

Consumers

Just as we thought we knew what organisations were for, we thought we knew what consumers were for. They were the consummation of the production process, the final link in the value chain. In an industrial organisation, raw materials came in at one end, to be worked on by labour and machinery, and finished products left from the other end to be bought by consumers. The car, television, fridge, telephone: all might be the preserve of the rich had we not created these value chains to make them on a mass scale and at low cost.[4] But these orderly value chains are now being scrambled up and, as a result, some consumers, in some areas of the economy, are playing very different roles. They are no longer passive recipients of goods delivered to them; they want to be participants in the creation of services they want.

A prime example of this participative consumer culture is World of Warcraft which has become, since its launch in 2004, one of the most successful computer games ever, with more than 8 million subscribers.[5] The average player spends more than 21 hours a week playing, and one study found that within eight months of the game's being released more than 15 per cent of the players had reached level 60, which would have required the equivalent of two months' non-stop playing.[6] Part of what makes World of Warcraft so attractive – critics would say addictive – is its social structure. Players create a game character, or avatar, which they guide through the game to acquire points, by killing monsters, completing quests and exploring unknown territory. There are 10 races, nine classes and two warring factions, Horde and Alliance. Players can talk only to members of their own faction; they can fight only the opposition. The action is split into several realms with up to 20,000 combatants in each. Each realm has a different kind of combat.

World of Warcraft was designed to be social and participative. Many tasks are too complex to be undertaken alone, putting a premium on collaboration among characters with complementary abilities. Soon after the game was launched, guilds started to form. Some, like the one founded by the Japanese Internet venture capitalist Joi Ito, have thousands of members and are like mini-societies with their own websites and private lore. Others have only a handful of casual members. Yet World of Warcraft is not a collaboration of altruistic hackers. The 8 million players pay a monthly subscription of $14.99. Extensions to the game have to be bought: the first to be released, World of Warcraft: The Burning Crusade, sold 2.4 million copies in the first 24 hours. And there is a burgeoning industry of 'real-money trades' whereby game workers, perhaps 100,000[7] to 500,000[8] of them in gaming sweatshops in China, accumulate points and currency to sell to players. In mid-2007 you could buy a highly skilled male night elf on the Aegwynn US Alliance server, complete with armour and other possessions, for $829.99.[9]

World of Warcraft announces the development of a mutant

consumer culture. The players are consumers: they buy the game's software, pay a subscription and purchase other goods and services. Yet Blizzard Entertainment, which makes the game's software, does not deliver a World of Warcraft experience down a production line. It gives the players the tools to create it among themselves. The consumers become participants: they put in a lot of their own effort, creating their own characters, and collaborate to create the social systems and battles that are the point of the game. Players of games like World of Warcraft work at their leisure. The World of Warcraft recipe – consume, participate, collaborate – scrambles up the cast-iron categories of the industrial era.[10]

The dichotomy of supply and demand does not make sense in a world in which products and services can be built by the people who consume them and where under the right conditions, demand can generate its own supply, unlocking a vast new source of value. Organisations built on high levels of member participation can have low costs: the most successful games companies sustain communities of millions of players with at most a few thousand employees. When consumers become participants, innovation becomes more dynamic and, in some respects, more manageable.

Most innovations fail because it is so hard to get inside the heads of potential consumers, to know whether they will really take up what seems like a promising new product.[11] Inventors are often very bad at guessing how consumers will use technologies. The inventors of the telephone thought that people would use it to listen to live performances on the London stage. They did not think it would be used for conversation. Innovation is so fraught with risk because the gap between a commercial company's idea of what consumers want and what they actually do want is often too large. Companies seek to close this gap through better market research, but rarely – if ever – completely succeed, and so long as the gap remains, so do the risks of innovation.

User-driven innovation, when consumers participate in designing new solutions, closes this gap: the consumers can design exactly what they want. A prime example is the emergence of the mountain bike, which was created when young cyclists in northern California started to take their bikes onto mountain tracks in search of new challenges. Traditional bikes were not designed for this terrain, so the rider-developers put together machines by mixing strong, old-fashioned bike frames, wider tyres to improve grip and drum brakes from motor cycles. For several years pro-am bikers built these 'clunkers' in their garages. Commercial manufacture began in about 1975, and a year later there were half a dozen specialist assemblers in the part of northern California that lent its name to Marin, the company that became one of the best-known mountain-bike producers. Almost a decade after clunkers first hit mountain tracks, the commercial bike manufacturers finally followed suit. By 2004, in the US mountain bikes and related equipment accounted for 65 per cent of all bike sales. A category that had been invented by passionate users was worth $58 billion dollars. The biggest disruptive innovation in modern biking history came from consumer innovators. Something similar has happened in other extreme sports, such as windsurfing and snowboarding. One study suggests that in active sports about 57 per cent of significant innovations have come from consumers.

Where technology has many possible applications, the users rather than the producers often work out what it is really for. Mobile-phone companies failed to predict that SMS messaging would become one of the main forms of communication among teenagers. Only when the technology fell into the hands of users did the full range of its application begin to emerge.

Consumer innovators are often also critical to disruptive innovations that upset an entire market. Mainstream companies working in mass markets often have powerful incentives not to innovate: they tend to reinforce past success and overlook small, emerging markets where there is little money to be made. In those marginal markets, passionate users, who are not interested in making a big profit, often carry innovation forward. Twenty years ago, no one in their right mind in a big record company would have dared to suggest promoting a form of music in which black men in inner-city ghettos expressed their anger at the world and glorified violence. Rap started as a pro-am activity, with people recording songs at home and distributing them on tapes. Twenty years later rap is one of the dominant forms of popular music in the world, touching many other aspects of popular culture.

Consumer innovators rarely work alone. They thrive in communities that freely share open-source designs. Newman Darby, who created one of the first windsurfers, published his designs in *Popular Science* magazine. Dimitry Milovich was granted a patent for his snowboard design in 1971 but made it clear he would never enforce it. John Dobson never patented the cheap digital telescope he created, so paving the way for a renaissance in amateur astronomy. The messages for mainstream companies are clear. Organisations built on knowledgeable and committed communities of users, like eBay, often find that good ideas come from the membership base who will also provide rapid feedback on whether a new service will work. When the link between the company and the community, the producer and the user, is tighter, innovation should be less fraught. If companies want to engage their consumers as innovators, they too will have to open up so their consumers can freely share ideas and modify their products.[12]

The spread of the web means more people will have the opportunity to be part of these sorts of communities. More of us, for more of the time, will be participants as well as recipients, players not just spectators. Cheaper and higher-quality technology puts powerful tools, which might once have been the preserve of professionals, into the hands of amateurs. Knowledge once held tightly by institutions is starting to flow into communities of pro-ams, breeding a mass do-it-yourself culture among amateurs who want to do things for themselves but to high standards.

People will participate in quite different ways.[13] At one end of the spectrum will be fans, who want to worship something created by someone else, whether that be a computer or a football club. To see a fan community at worship you have only to go to an Apple Store: it is almost a cathedral, a devotional space where people go for a rite of initiation into the Apple Way. The priesthood wander among the congregation who are sharing tips with one another and waiting for instruction from the evangelists. The most prolific fan community is the Trekkies, who worship the *Star Trek* series, attend conventions, dress up in character and make devotional offerings in the form of the more than 350 Trekkie-made feature films on the Internet.

Fans have no wish to create an alternative to the brand, team or product they worship; they want to claim their bit of it, to remake and add to it. Advertisers and marketers have always wanted people to become fans, which is why they aim to engage the consumer's imagination to make a product feel aspirational. The sociologist Erving Goffman, who studied advertising in the 1960s and 1970s, found that the most sophisticated advertisements were 'half-finished' frames which invited the consumer to fill in the remainder of the picture. Brands work by engaging the consumer's emotions, and that means getting them to make imaginative associations between the product and what they dream of. Latter-day user-generated content created around brands like Nokia and Apple takes this marketing strategy several steps further. Expect more efforts to be made to turn you into a fan.

At the other end of the spectrum of participation are the self-styled hackers – the digital craft aristocracy – who will build online communities that are self-governing, do-it-yourself and highly sceptical of commercial brands.[14] Where fans want their slice of something that is mainstream, hackers want to create an alternative to the mainstream. Hackers are suspicious of corporate brands and want to undertake self-governing craftwork.[15] They do not want to be managed, nor do they want to be commercial.

This is where it all starts to get rather confusing. More companies will try to follow Apple and persuade people to become fans of their products by claiming to have a hacker ethic beating inside them. Google is the prime exponent of this conjuring trick: never has the hacker spirit generated so much profit. Companies that create fans will, however, find them hacking their software and generating their own content and will be left wondering how to react. Lego, the Danish toy-brick company, found that a fan had hacked the software for its Mindstorms robots and made improvements that were proving popular with other players. Lego allowed the hacked software to flourish without endorsing it.

By the time the US Super Bowl came round in early 2007, the deliberate confusion of amateur and professional, consumer and participant, was near complete. Some of the most expensive advertising slots in history – $2.6 million for a 30-second broadcast to more than 90 million people – were given over to videos for Chevrolet, Alka-Seltzer, Doritos and the National Football League that were in part homemade. Doritos held a competition asking consumers to produce their own advertisement in return for free tickets and a $10,000 prize. One of the finalists, Weston Philips, estimated that his offering – which featured a leering Doritos-eating motorist – cost $12.79 to make. In 2007, amateurism was all the rage in mainstream media.

Nothing better embodies the confused character of participant consumerism than The Sims, the most successful computer game ever. Designed by Will Wright, perhaps the pre-eminent designer for mass participation, The Sims, with its add-ons, has since its launch in 2000 earned in the region of $1 billion for its publisher, Electonic Arts. Yet Wright estimates that player-developers have created at least 60 per cent of the content. They have even organised their own Mall of Sims, to sell their wares. More than 50 online shops – with names like Miles of Tiles, Synch 'r Swim and One Size Fits All – run by pro-am developers serve about 10,000 subscribers. The mall offers a plethora of objects for the Sims world: exteriors including Astroturf, fountains, swimming pools, tents, trees, shrubs or entire houses; interiors based on the work of real-world artists, from Salvador Dali to Andy Warhol. One religious-themed 'wall' depicting the crucifixion had by mid-2007 reached 1,364 downloads; walls with the *Lara Croft* and *Star Wars* themes had 2,195 and 2,488 downloads respectively. As Wright explained,

> Different games are competing for communities which in the long run will drive sales. Whichever game attracts the best

> community will enjoy the most success. What you can do to make the game more successful is not to make the game better but to make the community better.

The Sims is a platform that supports a vast, do-it-yourself community of gamers who develop and share their ideas. The point is not just to play the game but to add to it. What drives this economy, according to Wright, is peer recognition:

> The currency is exposure and recognition for their ideas. People are spending time creating cool objects – a lot of them are not spending so much time playing the game. How do we build the most thriving community online? We have to let the business model flow from that.

The Sims community built on the exchange of peer recognition nevertheless makes vast sums of money for Electronic Arts.

These bottom-heavy business models that rely on consumers' doing much of the work may, however, come of age in Asia, where people may not want to pay for expensive professional services. Timothy Chan knows how compelling these models could be in the developing world. I met him in a plush, private dining room in an exclusive restaurant designed for Shanghai's new rich, half-way up a skyscraper on a plot of land that in 1997 grew vegetables. In 1999 Chan left his job as a government adviser to start his own business, at the height of the dot.com boom. By 2004 Chan's company, Shanda, had 170 million registered users and 60 per cent of the Chinese online games market, with 10 million regular players. Shanda distributes its content across China by giving away the game's software, which it expects to be copied millions of times over. That means Shanda does not need a sales force or a marketing department. The players handle the distribution themselves. However, before someone can start playing a Shanda game they have to activate it, which means logging on to one of Shanda's servers, providing credit-card details or a pin number from a pre-paid card bought from a newsagent's. The games are huge social events. More than half the participants play from Internet cafés. The more players, the more action. Chan explained the appeal while crunching through a duck pancake:

> The average user plays for three or four hours a day. They concentrate on building up their character and profile in one game. More users means more distribution, which means more action, which attracts more players. It is a virtuous circle.

When a player runs into payment difficulties he loses access to his character. Then he has two options. He could buy a new character quite cheaply – but that would mean starting all over again, building up a new character. The alternative is to go to Shanghai, stay overnight in a hotel and queue outside Shanda's offices to reclaim the original character with all its accumulated history and skills. That costs 10 times more. Nevertheless, every morning about 500 people are in line outside the Shanda headquarters. To support this vast undertaking – 9,000 servers across China used by more than 170 million people – Shanda employs just 600 staff.

As consumption becomes more participative, at least when and where we want it to be so, the categories we use to carve up life will need to be rethought: for some people, leisure will be a form of work; in some fields, amateurs will be able to do jobs previously preserved for professionals; consumers will be able to produce; users will be innovators; demand will generate its own supply; communities, like those playing The Sims, will be a new source of wealth for companies.

Work

Seb Potter is a pasty, pale young man who spends too much time with his laptop. He programmed his first computer at the age of eight and has not stopped since. Sloping around his office in baggy jeans, T-shirt and trainers, Potter is part of a quiet revolution in how we approach work. He got involved in open-source software development as a student in 1998 when he helped to launch Evolt, an online community bringing together amateur and professional web-developers:

> We just exchanged emails and agreed it was a good thing to do. We divided up the tasks, depending on who could do what, and within a month we had a fully functioning website which people could visit to get tips and advice. We just wanted to make it easier for people to solve web-development problems they faced.

By 2004 the Evolt community had 7,000 members. As more people got involved, they asked new questions. The answers expanded the shared knowledge base. GetFrank, the web-development company for which Potter worked, encouraged him to spend a quarter of his time on open-source projects because that helped the company access software it could not afford to develop on its own. Potter's open-source work gave him a sense of personal satisfaction:

> I love problem-solving and if you are into software then pretty much the only way you can do that is in open-source projects

> because proprietary systems are closed. Open-source communities judge you on the ideas you have and the contributions you make, not on what you look like. If you have good ideas you get recognised. For me, work is the compromise. I feel most myself doing this kind of stuff.

Traditional companies find it difficult to tap into the passion and imagination of people like Seb Potter, which is why open-source communities could change the way we work even in organisations that do not make software. How can a group of 7,000 people share out tasks, build up a common knowledge base, develop a set of tools and provide valuable services without having an office, a management hierarchy, a human-resources department, a knowledge-management programme, 360-degree career reviews, corporate away days and embarrassing Christmas lunches with the boss? Collaborative approaches to work will spread because they expose just how dysfunctional and unpleasant it is to work for most large corporations.

Organisations exist to get work done. Any successful organisation must do three things well: it must motivate people to make the most effective contribution to the collective endeavour; must co-ordinate the contributions of many people to make sure the work gets done in the right order; and must innovate by encouraging people to learn and adapt. Sounds simple: motivate, co-ordinate and innovate.[16] Yet corporate organisations are in perpetual crisis because they find it so hard to do all three at the same time.

An organisation with a strong brand needs consistent products, services and processes often on a global scale. Wherever you go, Starbucks, Coke, Microsoft must mean the same thing. That can be guaranteed only if work is highly systematised. Yet all organisations face more intense challenges of innovation and adaptation as new technologies, competitors and consumer trends swirl around them. Innovating comes from bending or breaking rules, trying out new approaches, or at the very least delivering old ideas with a new twist.

Employees are pulled in different directions by the twin pressures to be efficient and innovative. People no longer work at the same time and in the same place. Workers can be deployed whenever and wherever they are needed. Performance management has become more insistent. Relationships at work have become more transactional and erratic. Loyalty and trust on both sides of the employment contract have withered. Management hierarchies are flatter, but that is not necessarily a recipe for more self-management, and disguises more centralisation of power in fewer hands. The angst this creates can be seen in corporate offices everywhere, and especially in the US.

Open and collaborative forms of organisation are emerging because they offer to resolve the tension between efficiency and innovation more effectively than the traditional corporation. Open-source communities seem to answer the three critical questions about work – how to motivate, how to co-ordinate and how to innovate – while requiring little in the way of top-down bureaucracy. They motivate a mass of contributors by providing interesting work, posing interesting questions and attracting interesting people to work with. The work is co-ordinated because the products clip together with modular architectures, performance is judged by your peers, and the community shares an overarching goal. Authority is exercised mainly by peer review and with a light touch. The traditional organisation co-ordinates work through a division of labour: people are allotted tasks divided up from above, on the assumption that the people who design the process or production line know what needs to be done. Open-source communities co-ordinate people as they innovate. People distribute themselves to the tasks they think need doing and for which they have the skills. A self-distribution of labour – if it works – is far cheaper and more innovative than a centrally planned division of labour.

By resurrecting the old-fashioned notion of craftsmanship for the digital age, We-Think's approach to work also speaks to younger generations who want to be in control of their lives, who want to extend their sense of authorship from how they design their living rooms to how they conduct their careers, to feel autonomous, take the initiative and be rewarded with a sense of achievement and recognition. Open-source programmers get recognition for the quality of their craft by participating in a community of their peers. We-Think gives people the chance to work in a community that can give them a sense of recognition and belonging: precisely what the modern corporation denies most people.

Open-source communities and other forms of We-Think provide an inspirational new model for how we can work together, collaboratively and creatively. Yet clearly this will not work in all settings. People have to be motivated to contribute, and there are plenty of tasks – such as collecting the community's rubbish – that most people will not willingly do out of a sense of curiosity. (That said, even rubbish collection is becoming more knowledge-intensive: recycling depends on motivating people to care about what they do with their waste.) Most people will continue to be paid to go to work in organisations. But We-Think's recipe will gradually infiltrate organisations to make them progressively more open and collaborative.

Some maverick firms have already begun to adopt these approaches. W. L. Gore, the maker of Gore-Tex, has sales of close on $1.4 billion but claims to have no managers, secretaries or even employees. Instead it has a global network of 6,000 associates, who jointly own the company. Salaries are decided collaboratively. Each new associate has three peer mentors who help to navigate his or her career. Even large companies are being drawn in this direction. In 1998, BT, the British telecommunications behemoth, created a Freedom to Choose scheme for field engineers, after an experiment with a particularly recalcitrant group of engineers in Cardiff. All attempts to get this group to work harder had failed, so in desperation the managers gave the engineers the right to self-schedule their work. Many of the decisions about who would work when were made in Internet chat rooms with the help of scheduling software. Engineers earned points by completing work, mentoring peers and leading teams. In March 2002 the pilot was extended to 20,000 engineers, who now self-schedule. After three years, the average engineer was earning more money and working two hours less per week. Productivity was up by 5 per cent and quality up by 8 per cent. BT is adopting elements of the open approach to work because that is the best way to motivate and co-ordinate knowledge workers at low cost.

The collaborative values of We-Think are also reshaping how places of work are designed. Modern organisations revolve around offices that are often designed to help managers to stay in charge. Offices are good for politics, flirting and gossip. They are dreadful places for intellectual curiosity. Creativity comes from diversity, from exposure to different points of view and experiences, but offices tend to make everyone conform to the corporate code. Innovation often comes from creative interaction with customers; offices are a good place to hide from the outside world. They also encourage territorialism – different departments on different floors – so it is difficult for people to cross boundaries to borrow and share ideas. Lateral or sideways thinking is virtually impossible in the standard office environment. The recognition that offices are dreadful for conversation and so for innovation has led many companies to try to become places that allow creative conversation, inviting social interactions that make it easy for people to talk to one another. Offices need a social milieu akin to that of a bustling city neighbourhood, where much of the life takes place on pavements and in cafés.[17] That is why people everywhere are turning offices into cafés: in the developed world the *latte* has become an essential work tool.

The call for more open and participative forms of work will sound Utopian to many and I might have been inclined to agree until I met Chris Sacca, one of the first employees at Google. Over dinner in late 2005, Sacca's account of how Google organised its work went something like this. Every Friday everyone in Google gathers for an all-company meeting – 7,000 people, face-to-face or connected by video. Anyone can ask the senior managers any question about policy, strategy or performance. Every week, each person working in one of Sacca's teams emails him five bullet points explaining what they have achieved that week and five more on what they plan to achieve in the week to come. It is up to people to sort out what they are doing without calling in a manager. Anyone can search through the bullet points submitted by anyone else, including those from the chief executive. Anyone in Google can launch a development project so long as they post the details on a central site so that everyone else can see what is going on. They can continue with their project until they want to recruit more than two people or they start to use some significant server capacity. Once their project has reached that stage they have to take the idea to a company council, which will make a decision about whether it should go ahead.

This all needs a hefty pinch of salt. Work cultures are rarely as open and democratic as people at the top of an organisation claim, even at a company as funky as Google. Yet if only half of Sacca's account is accurate, the company's work culture poses a huge challenge for traditional, top-down companies. An organisation that wants to match Google's capacity for innovation needs to match its work culture. Google is the most striking example of a company that has taken elements of We-Think-style work into the corporate world and created an extremely potent mutant: a money-making machine that espouses open-source values.

Of course it is ridiculous to imagine that most work could be like this in future. Call centres and retail outlets will be experience and service factories: highly regimented, delivering a high-quality, low-cost service. Trading floors in banks are financial-services factories. Even young scientists complain that lab work is highly repetitive and boring. Yet large organisations will increasingly feel the gravitational pull of open and participative ways of working. Traditional firms will have to become more democratic, open and egalitarian if they are to match the appeal of We-Think work. Edmund Phelps, Columbia University professor of economics, argued in his acceptance speech for the 2006 Nobel prize in economics that in the 20th century the focus in developed economies was on the satisfactions of consumption, but in this century the focus will be on providing satisfying work that involves solving problems, intellectual discovery, new challenges, personal growth and making a contribution that is valued by others.

Leadership

Leaders sit at the top of traditional hierarchical organisations, in the largest offices with the best views, because they have the special qualities and skills needed to make the decisions that count. Leaders give permission and approval. They make sure resources are 'aligned' behind their corporate strategies by creating the right incentives or issuing instructions. Decisions are passed up for approval; instructions and permissions filter down the chain of command. That's how most traditional organisations work.

But this closed model of leadership is increasingly ill-equipped to cope with the demands faced by large organisations. Organisations and their leaders need to be more alert and agile. Sprawling networked organisations can be thrown off balance by events in far-flung economies: British banks can be upended when lots of low-income households do not repay their mortgages in the US Midwest; the biggest oil company in the world can have its reputation tarnished by mistakes at a refinery in Alaska. Leaders in all walks of life operate in a far more open environment, under constant scrutiny from the media, regulators, their employees and now the wider world of citizen journalists, armed with their camera phones and blogs. Closed leadership is too slow for this world because too many decisions have to be passed upwards for approval.

Closed leadership is also out of kilter with the increasingly democratic ethos of the times. Authority is now more likely to be questioned and less likely to be meekly followed. Employees' support cannot be taken for granted; it needs to be renewed time and again. Nor is the desire for more open styles of leadership confined to a handful of trendy Californian companies. In 2003 an international consulting firm asked me to report on the views of emerging corporate leaders. From

My IDEA
OR Yours!
vino

South Korea and Brazil to Germany and the UK, the message was remarkably consistent: heroic top-down leadership did not work; people were no longer willing to be told what to do; young leaders were no longer willing to sacrifice everything else in their life, including their families, for their work. A young German manager put it this way:

> Once upon a time those above were nodded to, and up there someone sat all alone and decided and left their mark. Now the competencies are distributed on all levels, not equally, but at least so that one can more quickly decide. The trend is away from the authoritative style of management in which one person decided and others carried out. The emphasis now is more on teamwork.

We-Think plays straight into this desire for more open and accountable leadership. It is a myth that collaborative communities are wholly self-organising. All have founders and leaders. But their leadership is exercised in a way that is quite different from that of the traditional corporation.

This new breed of leaders makes very few decisions. Instead they focus on creating the norms and rules through which many other people can take responsibility for small parts of what the community does: leading a guild in World of Warcraft, looking after a module of an open-source programme, specialising in a particular part of the genome. They tend to be quiet, self-effacing and humble – the antithesis of the testosterone-charged, charismatic chief executive. They invite others to come forward with ideas rather than hogging the limelight themselves. Open leaders tend to come from and identify with the communities they lead: Craig Newmark, the founder of Craigslist, still calls himself a customer-service rep. Even if they are benevolent dictators, like Mark Shuttleworth, the founder of Ubuntu, they submit themselves to far more transparent and accountable decision-making. Leaders of mass collaborations, like Sydney Brenner, the architect of the worm project, attract collaborators to interesting questions, orchestrate creative conversations around those questions and embody the organisation's values in how they behave.

The more important innovation becomes and the more innovation involves combining the ideas of many people, both inside and outside an organisation, the more leaders will have to orchestrate creative conversations. In a creative conversation, each party must give something of themselves to encourage the others to reciprocate; must listen intently, not just speak their mind; must be prepared to adjust their ideas, not simply defend fixed views. Most managers are dreadful at creative conversations. Few have the time really to listen to people. Being a manager is about being in charge, knowing the answers, deciding what to do. That is why managers don't often have good conversations with the people they work with. They are too distant and their mere presence often kills conversation. (When I worked at the *Financial Times* we had a chief executive who had clearly been told to 'walk the floor' and chat to people from time to time. Every other Friday afternoon he would stick his head around the door to the editorial floor looking like a scared rabbit and every journalist in sight would hit the phones to avoiding making eye contact.)

An outstanding example of how the open-leadership style of We-Think communities can work in traditional organisations was Jorma Ollila, the long-time chairman of Nokia. This company became a leading force in mobile communications thanks to the entrepreneurial vision of a group of senior leaders, who charted it away from Wellington boots and cycle tyres into high technology when the market for mobile phones was still unproven. Yet Nokia is also an egalitarian and bottom-up company. At the corporate headquarters just outside Helsinki all Nokians have lunch together in three canteens on the ground floor. Everyone joins the same queue, eats the same food, shares the same space. Two-fifths of Nokia's 55,000 employees are in R & D activities. According to Ollila they are more like artists working in a creative community than engineers. When I met Ollila in Helsinki in 2005 I wondered how he led a community of 20,000 artists; this is how he explained it:

> Innovation and creativity are not individualistic. It's really about interaction, getting people to interact with one another in the right way. Leadership is about creating an atmosphere in which people get a kick from working with one another.

Ownership

Innovation thrives when ideas can be shared among consumers, developers, suppliers. When more people are involved in generating an idea, it becomes increasingly difficult to work out who did what and so who might own which bit. Collaborative innovation invariably requires a form of shared ownership. Yet this flies in the face of the conventional wisdom that private ownership – usually by companies with shareholders – is the best way to promote investment, productivity and innovation. When someone clearly owns something – such as their house – they tend to look after it better, invest more in it, care about it more than when it is something they share with others. Does We-Think imply a shift away from

you GET what you GIVE

shareholder-owned companies and the private property on which modern capitalism has been built?

The modern shareholder-owned company has related strengths that are hard to beat.[18] Companies have a remarkable capacity to adapt and evolve, and a company can outlive its founders and purpose. No one bats an eye about using Post-it notes made by a company that started life as Minnesota Mining and Materials. Companies have provided a vehicle for mobilising a mass of private investment into productive activities. Thanks to the innovation of limited liability and the stock market, investors can feel reasonably safe putting their money into companies that they cannot control and do not run. Shareholder ownership, meanwhile, puts managers under pressure to drive up productivity and to deliver financial results to shareholders. As a result, industrial companies have found more efficient ways to create the goods that underpin Western consumer wealth. Private ownership seems to create focused, efficient, well-capitalised organisations that are kept on their toes by their demanding but detached owners and can adapt to a changing environment.

Given this track record, it would be foolhardy to suggest that the investor-owned company should be replaced by the idealistic commune capitalism of open-source and We-Think, especially as co-operative organisations have such a chequered history. Common grazing land can be a platform for shared milk production; just as easily it can become a wasteland exhausted by over-use. Employee-owned companies and mutual societies, co-ops and communes have produced as many failures as successes. If a membership organisation grows, its membership can become too diverse and then it can find it difficult to take decisions. If the membership does not grow then such organisations can quickly become inward-looking and decline. Being a member of a co-operative brings costs as well as benefits: as a co-owner you have to devote time and effort to the community for little or no pay. Critics bemoan the short-term nature of relationships in flexible, networked companies. But many people would prefer this shallow commitment to the sacrifices required by life in a commune.

So any claims that common ownership has a bright new future should be treated with caution. Yet corporations seem more unstable than they were in the heyday of the industrial economy. The rate at which companies have left the Fortune 500 increased fourfold between 1970 and 1990. The shareholder-owned company emerged in tandem with the industrial economy in the late 19th and early 20th centuries. Modern developed economies depend on ideas and intellectual property as much as on industrial goods: in 1999 the US earned $37 billion from exports of intellectual property, compared with $29 billion from aircraft, its largest export-earning manufacturing industry. Shareholder ownership may have made sense for the ownership of steel mills, coal mines and sewing machines. It is less obviously the correct way to organise an economy based on the sharing of ideas. Even in the US, shareholder ownership has never been the only way people have owned productive enterprises together. All over the world there are employee-owned professional service firms; producer-owned farm co-operatives (which still account for 30 per cent of US agricultural output); consumer-owned energy and telephone utilities (in the US and elsewhere); occupant-owned condominiums, favoured among the American elderly; insurance companies owned by their policy-holders; and one of the fastest-growing sectors of the economy is non-profits – in health, education, day care and other services – organisations that often do not have beneficial owners but answer to a charitable or social purpose.[19] We-Think is adding to the already wide range of ownership options available.

Alternative forms of ownership and what economists call new business models often play a critical role in radical innovation, opening up new markets and creating new industries. When fertilisers were introduced to the US in the 19th century, farmers found it difficult to establish their quality and many were ripped off by unscrupulous suppliers. So farmers got together to make and distribute their own fertilisers through co-operatives. The US life-insurance industry was created in much the same way: consumers did not trust investor-owned companies not to run off with their money, so although several conservatively run private companies had been in the market for 30 years, by 1840 only a few hundred policies had been sold. In 1843, the first mutual life-insurer was formed as a co-operative of its policy-holders. By 1849 there were 17 mutuals and the market was reaching hundreds of thousands of policy-holders. Shareholder-owned companies largely withdrew from the market; many of the mutuals live on. The story of 19th-century farm fertilisers and life insurance is being repeated in the new markets being created by the Internet: alternative forms of ownership play a critical role in opening up new markets when people do not trust private-sector monopolists and can devise their own do-it-yourself solutions quite effectively. Whether these alternative forms of ownership last remains to be seen. In many areas of financial services, mutuals subsequently turned into shareholder-owned companies. But we certainly should not be surprised that radical innovation is coming from alternative social forms of ownership: it has happened before.

Whether collaborative and open forms of ownership of ideas such as open-source and Creative Commons will prosper depends on what happens to the law governing the ownership of ideas known as intellectual property. (Creative Commons is a form of copyright that allows users to share content very easily.) History, common sense and economic theory suggest that resources are most productively used when they are enclosed and owned by someone. The British enclosures in the 15th century turned common land into privately owned farmland and spurred a massive increase in agricultural productivity. Landowners had an incentive to invest in the land to make it more productive. Urban consumers got cheaper food as agricultural productivity increased. The peasants who used to graze their cattle on the common land lost out, but their descendants became farm-hands or factory workers.[20]

The example of the English countryside in the 15th century is now being applied to the 21st-century digital economy. The 'tragedy of the commons' argument is sanctioning the spread of private property into our shared intellectual and cultural life. A framework of legal fences, traps and gateways is being proposed in the US and Europe, backed by big companies. Patents are being stretched to apply to ideas, concepts, methods, collections of facts and other aspects of intellectual and cultural life that have hitherto been beyond the reach of private property. Intellectual property law started life as a way for business to protect itself against ideas being stolen. Patents were conceived as an orderly way to make knowledge available to a wider audience. These days big companies use thickets of patents to ward off competition and to make it difficult to find out how an innovation works. Intellectual property law is being distorted to the advantage of big corporations to stretch further, last longer and touch more ideas than ever.

Yet when shared resources are not grass and water but ideas and knowledge, they are infinitely expandable through collaborative innovation and so common ownership is often the best solution – as the designers of engines found in the Cornish tin mines in the 18th century. A massive extension of rigid definitions of private property into the intellectual and cultural arena will disrupt the delicate shared ecology of ideas on which all knowledge-creation depends. Innovation is invariably cumulative, the product of joint authorship, thriving on the interplay of different ideas, disciplines and viewpoints. News and cultural products, scientific theories and software code carry traces of other ideas and products from which they have borrowed and learned.[21] Private-property fundamentalism will destroy the shared basis for much innovation in science and culture.

Commune capitalism will not replace the shareholder-owned company. Yet open innovation blossoms with shared ownership of intellectual property, perhaps especially in emerging markets, and so shareholder-owned companies, out of their own self-interest, will increasingly have to find ways to share ideas with their partners, developers and consumers. We-Think will succeed not because it is noble, altruistic or morally uplifting but because it is the most effective way to organise mass innovation at scale. It works. As more shareholder-owned companies are drawn towards more open approaches to innovation, so they will have to experiment with more shared ownership of ideas.

We-Think is providing new ways for us to organise ourselves without necessarily having *an* organisation. The emerging highly organised activities do not have a single source of authority at the top. As no one seeks to be completely in control, people have to exercise responsibility themselves. We used to assume that complex tasks would get done only with a clear division of labour so that everyone knew whose job it was to do what and when. Yet in the new swarming organisations, people seem to distribute themselves to the work. The conventional wisdom is that consumers are happiest when they are waited on hand and foot. Yet in digital communes, consumers devote time, effort and imagination to make products for one another – for free. Instead of innovation coming from special people working in special places, in We-Think innovation is the cumulative work of multiple authors; it happens in many places, not just in specially designated zones of creativity and most often when a product gets into the hands of ingenious consumers. To somebody sitting in their office in a large media or software corporation, working their socks off to meet constantly updated plans imposed by managers who want them to deliver relentless growth with a smile, it must be slightly bewildering to find that their competition comes from a bunch of people who create products for free because they enjoy their work and no one is telling them what to do.

In the coming conflict between entrenched corporate organisation and open collaboration, what are we likely to see? There will be a much wider range of ways to organise ourselves, from Wikipedia at one end to *The Encyclopaedia Britannica* at the other; from Linux to Microsoft. There will be an intense battle between these open and closed models of doing business. Some businesses in music publishing and entertainment have found their older business models

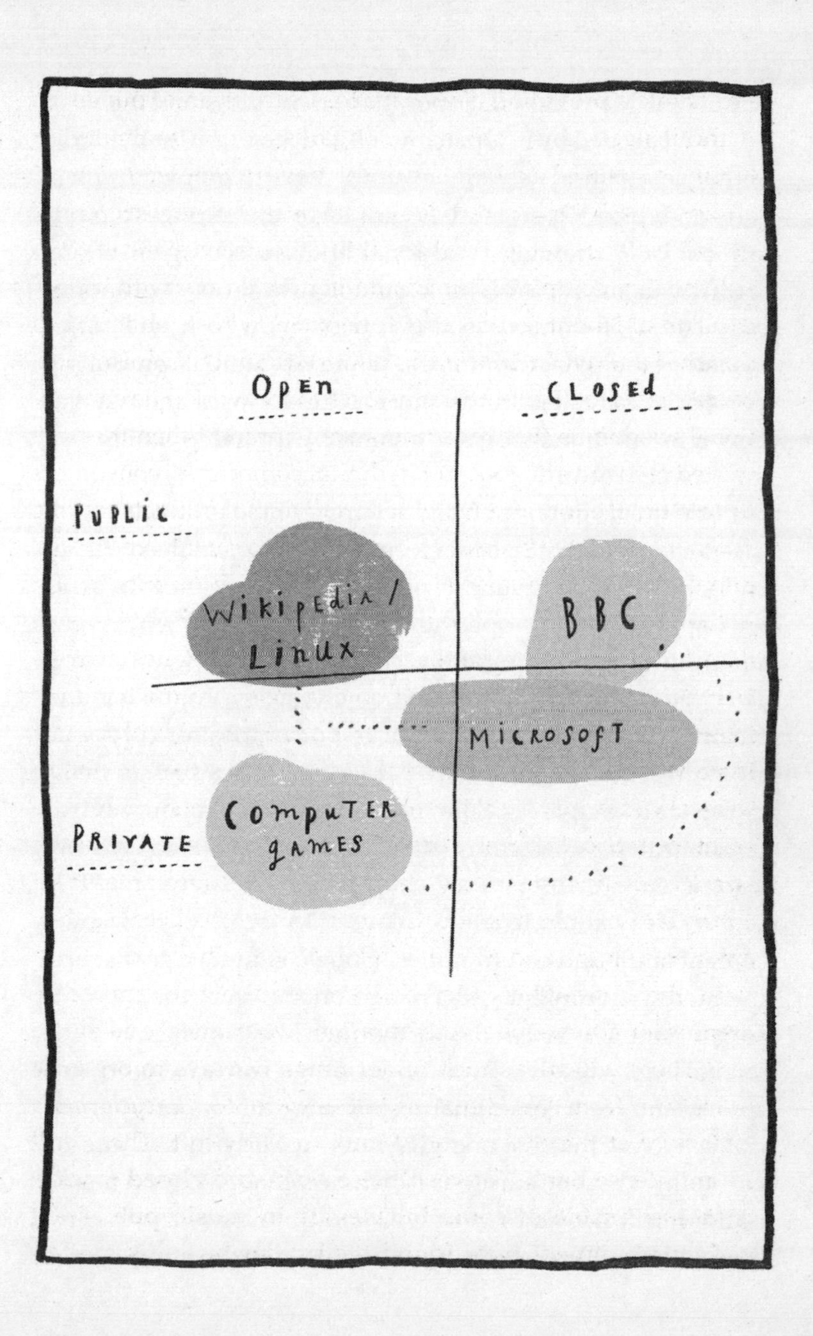
Open
CLOSEd
PUBLIC
PRIVATE
Wikipedia / Linux
BBC
MICROSOFT
Computer games

outmoded. Many closed corporations may withstand the siege but be weakened by it. Open models will be the most dynamic and agile, creating new and exciting ways to innovate, work, consume, lead. Open models will have the strongest gravitational pull, drawing traditional and closed organisations towards them. Yet We-Think cannot sustain itself economically based on voluntarism. If it tries to do so it will suffer the fate of the 1960s communes. While We-Think is producing powerful new collaborative approaches to work and innovation, it is based on models of common ownership that have a mixed track record.

Between the pure, open and voluntary models at one end of the spectrum and the classic closed corporation at the other, an enormous middle ground is opening up, where new hybrids will appear, mixing open and closed, public and private, community and corporation, collaboration and commerce. This middle ground will be extremely messy, confusing and creative. There will be plenty of opportunities for corporate cross-dressing as community-based businesses seek to make money and companies claim to be interested in community.

Computer-games companies such as Electronic Arts are leading cross-dressers: they charge for basic software while giving away tools to allow players to develop their own content and build communities. Google is in the same space: the more free content gets created on the web, the more its search engine is needed and the more links it has to mine for its search results. Even open-source software is in that space: Linux is a communal resource for a mass of commercial activity. Craigslist and eBay are similarly in this middle ground where community and commerce meet. Closed organisations will try, and many will fail, to become more open, while open collaborations will struggle to make money.

5 HOW FAR WILL WE-THINK SPREAD?

We-Think changes how we access and organise information and so it is bound to disrupt libraries and librarians, newspapers and journalists, music and book publishers. What is far less certain is how far it will spread beyond creative and cultural sectors in which information, knowledge and software play a critical role, into other aspects of life. Could it change how we learn, go on holiday, organise our personal finances, build houses, heat our homes, move around, eat, make fridges and cars or enjoy our leisure? We-Think might be having a big impact on Silicon Valley, Hollywood and New York's media businesses, but it is less central to the economies of Ohio and Iowa, Minnesota and Wyoming. In Europe, it is bound to have a bigger impact in the UK, which has a large services sector and a largely Internet-connected population, than in, say, Turkey, where close on 40 per cent of the population are employed in agriculture and manufacturing. It would be nonsense to claim that We-Think will revolutionise all economic sectors in every economy.

The creative and cultural sectors so far most directly affected – media and advertising, software and entertainment, film and television – account for about 7.3 per cent of the UK economy, employ perhaps 1.8 million people and are growing much faster than the economy as a whole. In the EU these industries account for about 3 per cent of GDP and in developing economies such as India and China the figure is closer to 1 per cent. We-Think could also have a big impact on some other service sectors, including communications (worth more than £30 billion a year in the UK) and publishing, which is counted as part of the manufacturing sector. In the UK and the US the sectors where We-Think will have a direct, early and high impact might account for perhaps 15 to 20 per cent of the economy.

In other fields only some activities might be affected. In general manufacturing, for example, the assembly line might not be much affected, but research, design, marketing and communications might well be. These industries will not be as fundamentally changed as music publishing and the media because their core product is neither information nor software. In the UK, these medium-impact sectors include education (£61 billion), health (£75 billion), public administration (£55 billion), retail trades (more than £100 billion), financial services (£86 billion) and business services (£55 billion). Taken together, these medium-impact sectors are probably worth about 50 per cent of the economy.

Then there are sectors in which digital technologies, information and software play a very small role, where We-Think will have little or no direct impact, certainly in the short run: agriculture, quarrying, mining, oil and natural gas, the £22 billion-a-year food-processing industry. These low-impact sectors make up about 30 per cent of the UK economy.

These proportions will differ in different economies around the world. The more developed an economy's creative and cultural sectors, the larger the media and publishing businesses and the bigger the services sector, the more impact

We-Think will have. In economies where basic manufacturing, extractive industries and agriculture are larger, We-Think will have less of an impact, at least in the short term.

A different way to estimate the likely impact is to examine which professions might be affected. We-Think may have a big impact on the 185,000 designers in the UK, the 9,000-plus librarians, the more than 35,000 journalists and the 300,000 software programmers. It will have some impact on the 435,000 teachers, 240,000 doctors, 800,000 engineers, the 3 million who work in retail sales and the 4 million managers. Far less directly affected may be the 480,000 who still work in agriculture, or the hundreds of thousands who provide personal care services.

Another way to gauge the potential impact is to examine what households spend their money on. The Institute of Fiscal Studies estimates that between 1975 and 1999 the share of expenditure on 'bread-and-butter' items declined from 40 per cent to 27 per cent of consumer spending in the UK, while the proportion spent on services rose from 29 per cent to 42 per cent, with expenditure on leisure goods increasing by 93 per cent and on entertainment by 109 per cent. In 1997/98 British households for the first time spent more on leisure than on food. In the following 10 years the share of household income spent on cinema, video games, music, satellite and cable subscriptions, Internet access and DVDs more than doubled. In the developed economies people will be spending more and more on the kinds of activities that We-Think might affect. Even so, a lot may go untouched: Britons spend close on £120 billion a year on fuel and other basic household expenses that may be largely unaffected by these developments in the digital economy.

In the short run it seems We-Think might have a direct and

disruptive impact on, at most, 20 per cent of the developed economies – on knowledge- and information-based services, creative and cultural industries, media and communications, where it is likely to change how goods and services are produced and consumed. Sectors such as education, business services, financial services, health, some manufacturing-related activities and some aspects of retailing will be affected but less fundamentally. These medium-impact sectors account for perhaps 50 per cent of the economy. Finally, there will be sectors worth about 30 per cent of the economy – agriculture, mining, quarrying, basic energy production, utilities such as water, and basic personal services – where the impact of We-Think and the web will be very limited.

These are crude estimates and we are at only the earliest stages of finding out how We-Think could change our culture and economy. Much will depend on how the web's collaborative potential unfolds. Let's examine how the potential for participative and collaborative approaches might change manufacturing, scientific research and public services.

From We-Think to We-Make?

Rob McKewan has become something of a legend among advocates of open-source business methods. The former chairman and chief executive of Goldcorp Inc., the Canadian mining company, was so frustrated by his geologists' inability to pin-point substantial gold reserves at his mine at Red Lake, Ontario, that he decided to post all the company's proprietary data – 50 years' worth of surveys, maps and geologists' reports – on the Internet to see whether the global community of geologists could locate the gold. The lure, apart from the intellectual challenge, was $500,000 of prize money. McKewan wanted to open up the data to test his own team's recommendations about where to drill for gold, and he also wanted to see if the outsiders could come up with ideas his researchers had not thought of. The challenge attracted 1,400 participants, many of them geologists but also physicists, mathematicians, complex-systems specialists, people with skills Goldcorp did not have. Eventually 140 of these made submissions and half of the 28 winners identified drilling sites that insiders had not spotted. Red Lake, which had been an industry laggard, became one of the richest gold mines in the world, with 6.6 million ounces of proven reserves.[1]

Goldcorp's story shows that open-source styles of working can be applied to aspects of the most basic industries. How far can similar recipes be applied to manufacturing? Highly collaborative approaches to draw together many independent contributors might make sense for writing software, which is easy to replicate, modify and distribute, but making physical goods requires investment in machinery and raw materials, factories and fork-lift trucks, which are not easy to reconfigure. To recoup such investments hardware manufacturers need to make a profit on products; they cannot share them out like gifts. Do open-source methods make sense as a way to make cars, planes, fridges and other physical goods? The answer, tentatively, is yes. In time We-Think could pave the way for We-Make.

Manufactured products have to be designed, and a design, like a software program, is a piece of information. Design and engineering have long been based on collaborative practices, from shipbuilders in the 17th century to the clusters of Silicon Valley. The Open Architecture Network (OAN) is a virtual space in which architects post their drawings for the public to examine, download and build. By June 2007 the network had 400 designs, 40 chapters of architects around the world and almost 6,000 contributing members. The OAN runs open design competitions on challenges facing the world's poor, such as the quest for low-cost clean water and sewerage systems, and provides up to $250,000 for the implementation of a winning design. OAN designs include The Clean Hub, a solar-powered water harvester, electricity generator and composting toilet that costs $15,000 to install, and the In(out) side House, an environmentally-friendly family house that would cost about $125,000 to build.

A site called Instructables – set up by Squid Labs, a design and innovation agency – is becoming an online directory of DIY design projects, an eclectic fusion of a home-improvement magazine and a how-to-build-it book from the 1950s – except that much of it appears to have been written by people with access to industrial lasers. The projects range from simple products, like a hammock that can be made from a single sheet of material, to complex pieces of electrical engineering, such as a touch-sensitive display like that used on the Apple iPhone. There is even an Instructables guide to building your own jet engine. One of the most developed Instructables projects is Open Prosthetics, created by an engineer who lost an arm in the Iraq War, which allows people with artificial limbs to share the adaptations they have designed. It includes a section called Pimp My Arm.

A site called Crowdspirit is aiming to apply similar open design principles to standard household and electrical goods such as CD players. Meanwhile Charles Collis, a research engineer at Dyson, the innovative British household-goods company, is pioneering open-source software for computer-aided design using a public library of component designs. Collis imagines 'an Internet of physical objects' – designs that can be shared, amended and passed around as easily as emails. The Society for Sustainable Mobility is orchestrating an international design collective to create an open-source car: a green SUV with a range of 600 miles which might go into production in 2011 at a price of $25,000. The designs are open – the public can examine them – but the rights to manufacture the eventual product will be more closely controlled to attract private investment.

None of this is likely to ring alarm bells among the world's major manufacturers – Matsushita and Siemens, GM and Toyota. But in the not-too-distant future there will be public libraries of open-source designs for a wide variety of basic products. An owner of an Apple computer can use software called GarageBand to create their own music, mixing melodies, styles and rhythms stored in the computer's memory. Soon we could have GarageFactory which would allow someone to do the same for objects. It is not a big step to imagine that the design could be sent electronically to a manufacturer in Poland or China who, armed with computer-controlled machine tools, could make the product. A service called Spreadshirt already allows people to do that with T-shirt designs. In future the same will be possible for other products.

More opportunities for We-Think will be opened up by the inclusion in physical objects of software and electronics, which already account for 30 per cent of the value of the average car. As of 2007, the web's collaborative culture is confined mainly to computers, because they are still the main products that carry software. When many more physical products become mini-computers – fridges, televisions, cars, phones, houses – then the file-sharing and peer-to-peer trends affecting computing will spread to other products. In the early 1900s when cars were first mass-produced, middle-class motorists added features to basic cars including radiator hoods, safety-belts, interior heaters, boots and reclining seats and shared their tips in 'how to' motoring magazines. That spirit of amateur tinkering and innovation may yet return to the car through its software and entertainment.[2]

Something like that may happen to mobile phones. In May 2006 in Silicon Valley the Homebrew Mobile Phone Club was launched to create open-source mobile phones so that people could design their own applications. A prototype for the first Linux-based phone, the Tuxphone, made from standard components, came out in early 2006. In March 2007, First International Computer (FIC), Inc., a Chinese manufacturer, started OpenMoko, a project to create open-source phone software. FIC will not need its own software-development programme if it can organise a user community to help it. The fully open-source mobile phone is some way off; the capability to make it, however, is starting to emerge.

All of this will make even more sense as manufacturing becomes more networked. Manufacturers increasingly assemble sub-systems made and designed by suppliers to be clicked together like Lego bricks. Design innovation for components takes place among networks of collaborating suppliers of these modules. Just as the Cornish tin mines created a generic Cornish engine in the 19th century based on their shared designs, so China's manufacturers are doing something similar with basic products such as motor cycles. In the 10 years from 1995, Chinese motor-cycle production grew from 5 million to 15 million a year. By 2007 the Chinese industry accounted for half of global motor-cycle production. The industry, which got going by copying Japanese designs, has since started to innovate, with highly collaborative suppliers in Chongquing and Zhejiang sharing information in dense networks. Chinese bikes are built with modular designs that mean innovation in a single sub-system – like the brakes – will not affect the design for the whole bike. Suppliers of adjacent parts work with one another to make sure their components fit together. A delicate mix of competition and co-operation has speeded innovation and cut costs: the price of the average Chinese motor cycle made for export has fallen from $700 to $200.[3]

Even more ambitious efforts to create widely distributed manufacturing systems are just at an early stage of development. Do-it-yourself, collaborative and small-scale manufacturing might become economic, if designs can be downloaded for free, machinery becomes as cheap and easy to use as a computer and raw materials can be easily obtained. Manufacturing needs to be centralised and highly capitalised to reap economies of scale for complex goods like cars and fine chemicals. But for simpler, basic goods – tin roofs, plates, combs, stoves, spades, buckets – manufacturing might become much more localised. Just as the fax machine, printer and photocopier have spread the world over, so could low-cost manufacturing of machines that make reliable products, customised to local needs.

One such machine could be based on Bath University's RepRap, which looks like a large photocopier and can make three-dimensional objects from designs stored inside its computer. In the 1950s the mathematician John von Neumann imagined a universal constructor: a computer linked to a manufacturing robot that could make virtually any physical object, including replicating itself. The closest the world got to such a machine was the replicator in *Star Wars*, which could make any object out of thin air; the RepRap might make that a reality. The RepRap developed as a rapid prototyping machine, making models that are built up by thin layers of

How do WE STOP THEM BREEDING...?
RAPIDREPLICATOR

plastic being sprayed on top of one another. Adrian Bowyer, who leads the project, has devised a way to insert electrical circuits, motors and sensors as well as moving parts into the models. High-end rapid prototyping machines used in design studios to mock up models, such as concept cars, cost about $25,000. Bowyer hopes his machine will cost about $400. A poor village equipped with a shared RepRap could make plates, roofs, stoves, lights, tables and cutlery so as long as it had the basic raw materials and could download open-source designs. Bowyer imagines RepRap machines replicating themselves, building more machines to make more products, in time allowing production of even complex products like drugs to become more localised.

Neil Gershenfeld, a visionary scientist at the Center for Bits and Atoms at MIT, has developed something very similar, an integrated manufacturing centre called a Fab Lab. Each lab is a suite of industrial fabrication tools, including laser cutters, that make three-dimensional objects, as well as machines that make antennas, circuits and other precision parts. Larry Sass, an MIT architecture professor, has adapted the Fab Lab to make a plywood house for about $2,000. Gershenfeld predicts the lab's costs will fall from $25,000 in 2007 to below $10,000 within a decade. In Ghana a Fab Lab is making mobile refrigerators, and in India another is producing gears to bring ageing copying machines back to life. Gershenfeld argues Fab Labs will allow people to make a much wider variety of products tailored to their local needs – niche manufacturing – and so sustain local innovation and entrepreneurship. Fab Labs could democratise how we make products, just as computers are democratising how we access information. Fabricators may migrate into our homes just as computers and printers have. If Gershenfeld is right, eventually we may be able to manipulate, redesign and make objects in the way we edit and redesign PowerPoint presentations.

The idea of swarms of rapid replicating machines, making products using a treasure trove of open-source designs, sounds far-fetched. Yet each of these developments is an extension of an already established trend. Close-knit communities of engineers have shared designs for centuries; the web makes this sharing much easier on a much larger scale. The spread of software and computing into a much wider range of products means that hackers will have a much wider range of objects to tinker with, adapt and improve. More companies may find themselves in a position similar to Lego's when a user community of almost 900 developers formed around its reprogrammable Lego Mindstorms robots within weeks of the launch. Bowyer and Gershenfeld may sound mad, predicting a world of near limitless manufacturing capacity in which all matter could be reprogrammed to make virtually anything. But in one sense they are simply proposing an addition to the capital stock of any average middle-class household that already owns a Black and Decker drill and workbench.

Yet even if distributed manufacturing is technically possible, it is doubtful it will make economic sense for many consumers. It seems unlikely that networks of rapid replicator machines will replace large factories, chemical plants and production lines. Too much investment and learning has been sunk into the industrial infrastructure of the developed world to make it vulnerable to that kind of upheaval. In the rich world Gershenfeld and Bowyer's machines will be mainly used by DIY fanatics and craft businesses. It is unlikely many consumers will turn away from high quality branded goods made by Sony, Nokia, Toshiba or BMW in favour of a bit of DIY that may or may not work.

However in the developing world, where most manufacturing will be done in future, these ideas could have a dramatic impact. Western consumers may scoff at do-it-yourself manufacturing but in a poor African village, miles from the nearest city, something like a rapid replicating machine could make much-needed basic tools: cooking utensils, pumps, stoves and other household goods. More Chinese manufacturers might follow the lead of the First International Computer company, by sponsoring a user community in the developed world to create open-source designs for products to be made at low cost in China. In time, the way things are made will change: We-Think really *will* pave the way for more We-Make.

Public services 2.0

Lynne Brindley's predecessors in her role as chief executive of the British Library had the daunting job of organising, conserving and making available the library's vast, ancient collection. But they did not have to face Brindley's challenge: the possibility that in the era of Wikipedia and Google the traditional library might be outmoded. Libraries provide orderly access to information thanks to a classification system that few non-librarians can fathom. Until recently that order was maintained by a relatively straightforward process: a manuscript would be published by one of a relatively small number of companies and a copy deposited with the library, thanks to legislation that imposes that obligation on publishers. The publication would be classified, marked up and stored on shelves until someone ordered it. Librarians sat on top of this system, surveying the flow of information in and information out.

Public libraries are model civic institutions: open to all, freely accessible, they exist to serve citizens rather than the government and do not seek to make a profit. Many other public institutions – schools, parks, the BBC – take their lead from the open ethos the public library created. So the crisis of identity and purpose afflicting public libraries is unsettling the foundations of the public realm. It is a sign of what might be to come for other public services.

Brindley is troubled by three dilemmas. What should a library collect in an era in which virtually everyone with an Internet connection and a computer can become a publisher? The challenge of writing history used to be to find and then make sense of what little documentary material was available on scraps of paper, such as long-lost letters. Historians worked imaginatively to fill in the gaps in what they could find. Yet when historians come to write the history of every era from now on, they will have to choose what to examine from a bewildering array of material, much of it published informally, frequently without peer review, and more often than not in formats that make it difficult to store: weblogs, videos, PowerPoint presentations, podcasts, Facebook profiles. Brindley says her job is like trying to 'organise a large bowl of spaghetti'. Does a library collect all of the web's content worldwide, something that can be done only from time to time? Or does a library focus on just bits of the web, and collect those regularly – in which case, what bits does it focus on?

How do you make this much larger range of material available to a library's users? If libraries are threatened on one side by user-generated information spaghetti, on the other is the looming presence of Google, which in 2006 announced it would digitise and make available free of charge 20 million out-of-copyright books held by a group of major libraries. Not everyone can make it to the British Library on London's Euston Road during opening hours; but thanks to Google

INFORMATION
S

they will be able to access many books free, at any time of day. One measure of the demand Google will unlock was the University of Michigan Library's experiment with 10,000 monographs that were idling on the shelves, rarely used by the university's population of 40,000 students and faculty. When these monographs were made available on the web to a potential readership of hundreds of millions they got between 500,000 and 1 million hits per month.[4] The threat from Google's digital archive is forcing public libraries into unusual alliances. Digitising books and running servers costs money that few public libraries have. The British Library has teamed up with Microsoft to digitise 100,000 books. In future large public libraries may become public-private ventures.

Yet at the same time libraries may become more dispersed, community ventures because library users will store more content on their own computers. It is not hard to imagine digital libraries as file-sharing systems – a bit like Kazaa, the music file-sharing system, but for the written word – where people find books on the hard drives of other people's computers. It is difficult to knock on someone's door and ask to borrow a book from their shelves; borrowing will be much easier when books are digital and people allow one another access to their public material.[5]

Communities may blossom around these dispersed digital collections, and pro-am librarians may take much of the responsibility for finding and filing content. Brindley foresees a 'rebalancing' of the relationship between librarians and users:

> It is going to become much more like a relationship of equals. One possible model for the library is to mix the best of the traditional – we take very seriously our mission to conserve the

> national heritage and make it publicly available – with growing levels of user participation and our facilitating communities of practice that grow especially around digital content.

In 2020 the British Library will still be a building on London's Euston Road, although by then it might be known as perhaps the Cisco British Library. The Library will prosper only by making available a much larger range of material in digital form, much of it stored on computers owned by other people and organised by communities of users working with librarians. The library of the future will be a platform for participation and collaboration, with users increasingly sharing information among themselves as well as drawing on the library's resources.

Libraries are model public institutions, open, civic, tolerant, providing a public platform where people can pursue a bewildering variety of private hobbies and interests. Could We-Think transform other public services as much?

Public spending on education, defence, policing, social welfare and health accounts for a low of 28 per cent of GDP in South Korea, 36 per cent in the US, 45 per cent in the UK and more in many other European economies, with highs of more than 55 per cent in Scandinavia. The web's impact on society will be even greater if it touches these services, which in most societies are financed and provided collectively. It is not hard to mock the very suggestion: hospital operations bought and sold on eBay, an army organised through Facebook, an education system that relies on Wikipedia, pro-am policemen. We will not do away with schools and teachers, hospitals and doctors, police officers and their stations or with professional armies. We-Think currently thrives on voluntarism. The state often forces us to be prudent: paying taxes, attending school, saving for our future, cutting our carbon consumption – activities we might not see the point of or for which we might not otherwise be able to summon the will. Yet the We-Think culture of participation and collaboration will seep into these areas because public services are in need of radical renewal.

Post-war public services were founded on a paternalistic ethic of professional control and expertise. Professionals delivered solutions to people in need. If a society wanted more education it employed more teachers and built more schools. More health meant more doctors and hospitals. Safer streets meant more police. Yet public schools and hospitals, welfare and waste systems now seem designed for an era when people were more deferential, when the economy, social life, jobs and families were more stable. Many public goods – like a safer, healthier community – cannot be delivered as if they were a parcel on DHL. Complex public goods have to be created from within towns and cities, often through collaborative innovation that involves public servants, private entrepreneurs, households and voluntary groups. Public-service bureaucracies are often too clumsy and cumbersome to respond people's demand for more personalised services. These institutions are deeply entrenched, resistant to reform, closed to new entrants and slow to embrace new ideas. Top-down, target-driven reforms to public services are yielding diminishing returns even when they are accompanied by very large increases in public spending: in the UK health spending doubled between 1997 and 2007, much of it to employ more nurses and doctors, but satisfaction with the health service did not double and the nation did not get markedly healthier. That is why more radical approaches are needed, to harness collective self-help and mass participation – the We-Think recipe. Take education as an example.

Schools, or something like them, will remain central to education and there will be a continued push to deliver higher standards, more consistently, for all pupils, with better teaching, in better classrooms. Yet schools are out of kilter with the world children are growing up in. In a world in which everything seems to be 24/7 and on demand, schools operate with rigid years, grades, terms and timetables. That might have made sense when most people worked at the same time, many of them in the same place, on the same tasks, their lives organised by the factory siren. But people increasingly work at different times and in different places. Schools are factories for learning in an economy in which innovation will bc critical. Schools do not need just to be open for longer, more flexible hours. Traditional schools do too little to encourage individual initiative and collaborative problem-solving; learning is cut off from real-world experiences; teaching focuses too much on cognitive skills and too little on the soft skills of sociability, teamwork and mutual respect. Nor are schools necessarily the most important places where children learn. Families are as important to education as school. An integrated education policy would focus on how schools interact with families, including learning supports at home, working on raising family aspirations for learning. Children spend 85 per cent of their waking hours outside school: increasingly they learn from the games they play on computers and from the television. They organise their lives through their mobile phones and social networks.

One way to look at the challenge facing the modern education system is as a computer game with a million players that needs only 1 per cent to become player-developers, creating content for other players, and that has an unpaid development workforce of 10,000 people. If we could get 1 per cent of the 7 million or so pupils in British state education to be player-developers that would be about 70,000 children, or 20 per cent of the teacher workforce. Some schools are already going down this path. In the late 1990s Steve Baker, the principal of the Lipson Community College in Plymouth, found himself wrestling with how to motivate children to see learning as something they had a stake in. Baker decided to tailor more learning to the needs and aspirations of different children. Yet he had no money to employ more teachers. The solution was to make learning a more peer-to-peer activity. Older and more able children at Lipson become lead learners to help other children learn. In computer and maths classes at Lipson it is common to find a child helping two or three peers through a problem. Baker flipped the school on its head and saw the children as part of the school's productive resources, not just as its consumers. The lead learners benefit because they get additional recognition for their contributions and their own learning is reinforced by their having to explain what they know to others.

Imagine how a town might apply the logic of We-Think to education. An eBay for learning, a city-wide learning exchange, could match learners with people with the skills to teach. If someone needed a tutorial in using GarageBand software, not something taught at school, they could find someone with the skills to teach them. More learning opportunities would be modelled on large-scale, multi-player games in which players discover challenges and acquire the tools and skills to overcome them together – for example, a city-wide sustainability challenge using maths and science skills. A video site called YouLearn would encourage pupils to post their own solutions to questions, complete with user ratings and comments. A city-based Wikipedia-style resource of facts, figures, information and insight would be created by and for the city's citizens and for its children. Tools would allow one generation of learners to follow in the footsteps of others, learning from their mistakes. Social networks could link people so they could learn with and from their friends, online and offline, in coffee shops and homes.

At the moment education is schooling, a special activity that takes place in special places at special times, in which children are largely instructed in subjects for reasons they little understand. A We-Think approach would offer learning all over, all the time, in a wide variety of settings, from a wide range of people, more tailored to small groups and individual needs. Pupils would have more say and more choice over what they could learn, how, where and when, from teachers, other adults and their peers.

Will Wright, the designer of The Sims, who knows how to get millions of teenagers to engage in an activity, told the *New Yorker* magazine what he would do with education:

> The problem with our education system is we've taken this kind of narrow, reductionist, Aristotelian approach to learning. It's not designed for experimenting with complex systems and navigating your way through them in an intuitive way, which is what games teach. It's not really designed for failure which is also something games teach. I mean, I think failure is a better teacher than success. Trial and error, reverse-engineering stuff in your mind, all the ways that kids interact with games – that's the kind of thinking schools should be teaching. And I would argue that as the world becomes more complex, and as outcomes become less about success or failure, games are better at preparing you.

Or, as Illich put it in *Deschooling Society* in 1972, 'Good institutions encourage self-assembly, re-use and repair. They do not just serve people but create capabilities in people, support initiative rather than supplant it.'

The We-Think recipe of participation and collaboration could also be applied to healthcare. In the developed world, the assumption is that hospitals and doctors deliver health: the patient goes in at one end ill, is worked on by doctors and nurses, and emerges out at the other, like a finished product, well again. The hospital-based healthcare system grew in response to the spread of contagious diseases accelerated by urbanisation and industrialisation in the late 19th century. Now in the US and the UK this system is dealing with an epidemic of chronic disease mainly due to people living longer, exercising less, and eating and drinking more.

In the UK, 45 per cent of the adult population have at least one long-standing medical condition. Amongst the fastest-growing group of the population, those more than 75 years old, the figure is 75 per cent. By 2030 the proportion of 65-year-olds with a long-term condition will double. Chief among these conditions is diabetes: more than 2 million people in Britain are diagnosed diabetics and a further 1 million are diabetic without realising it. If type 2 diabetes, the kind linked to lifestyle, is caught early, it can be kept in check. Yet between 40 per cent and 50 per cent of diabetes cases are not diagnosed until it is too late. Then people become dependent on having regular insulin injections, which in the UK involves repeat visits to the doctor. Diabetes, a preventable and manageable condition, costs the NHS £5 million a day – 5 per cent of total expenditure and 10 per cent of hospital in-patient costs. The hospital-based health system, with its heavy fixed costs for buildings and professional staff, is being clogged up by people with conditions that need to be prevented, managed and treated at home or in the community. Our hospital systems will become unclogged and health will improve in the long run only if patients become participants, producers of their own health, looking after themselves more effectively and relying on doctors less.

New generations of intelligent sensors and monitors will allow many of the tests that GPs now do to be done at home, with more self-assessment and self-diagnosis. As an example of the potential, take an NHS thrombosis-prevention service in north London that serves 5,000 patients taking drugs that reduce the risk of their blood's clotting. The patients go to their local GP surgery each week for a blood test, which is sent to a central unit for assessment. If the GP tests are done by 11 a.m., then the results are back by 1.30 p.m. and the unit writes to people who need to change their dosage. If it is urgent, they telephone. In Germany, however, similar patients do the tests themselves, whenever they like, analyse the results and change their dosage accordingly, using a small machine that costs about £400. The north-London unit employs scores of nurses to do tests that could easily be done by the patients if they had the tools, the skills and the self-confidence. Finding solutions like this by motivating and equipping people to participate in their own healthcare offers the biggest opportunities for productivity gains in the future.

People with long-term conditions need help, advice, support and tools close to hand, without having to visit a doctor: that means more organised peer-to-peer support. Already, articulate and well-run communities of patients and carers are organised around long-term conditions like Alzheimer's. Imagine a health system where many of the basic tests are carried out at home or at a nearby pharmacy; where much of the information you need in order to understand your condition is available on the web, organised by a competent pro-am community; and where the best way to learn how to cope with the non-medical aspects of a condition will be to join a social network of peer support.

We have been schooled to regard health as a service delivered to us, when it should primarily be a responsibility we all exercise. In *The Limits to Medicine* Ivan Illich described a health system based on personal responsibility rather than service:

> Success in this personal task is in large part the result of the self-awareness, self-discipline, and inner resources by which each person regulates his own daily rhythm and actions, his diet and sexual activity ... The level of public health corresponds to the degree to which the means and responsibility for coping with illness are distributed among the total population.

These ideas are not appropriate to every public service. The idea of self-help can be exploited to get users to do more of the work. People will continue to want to be consumers some of the time rather than participating, which can sound like hard work. Some collective public services do not allow for much participation – mass-transit systems for example – although even defence and security ultimately rely on citizen participation. The public sector nevertheless will have to adjust to the demands of a population that has grown up with MySpace and Facebook, Wikipedia and eBay. More public organisations are using internal wikis to organise information from many sources, from the US Office of the Director of National Intelligence to the Environmental Protection Agency. Younger generations will expect more say and choice, more opportunities to participate and collaborate.

Pioneers are starting to use the social-networking web for public good. In some cities in the UK, citizens can add information to online maps to alert councils to dumped cars and rubbish that needs collecting. A US scheme, Zipcar, gives people a share in ownership of pools of cars in cities across the country, allowing them to get access to a car just when they need it rather than have it sitting in the garage most of the year. Another scheme, GoLoco, aims to use the power of social networking to revive the flagging culture of sharing cars for commuting. At the moment we have just either very public forms of mass transit – buses and trains – or private cars and cycles. Zipcar and GoLoco's approach, allowing people to make flexible use of shared transport resources, will become more attractive as more US cities introduce congestion charges to reduce car usage. In the Netherlands, a police inspector in Utrecht has created a system for citizens to help the police in solving crimes, not unlike the approach Rob McKewan took at Goldcorp, and a social-networking site has been created to help people look after one another's ageing parents: you can sign up to look up someone's parents in Rotterdam and someone else in the network will reciprocate by looking in on yours in Maastricht. Public services are most effective when they mobilise citizens to make their own contributions alongside those of professionals.

A public sector that does not create platforms for its citizens to create solutions for themselves, together, will soon start to seem old, outdated and tired. It will also be far less productive and effective in creating public goods. The future of public services rests on their becoming platforms for participation and collaboration, mobilising citizens as player-developers in creating public goods.

Shared science

In popular mythology, science is a lonely activity, undertaken by eccentric boffins in dark laboratories late at night. Mounting pressures to commercialise science are pushing more scientists to patent their discoveries before sharing their results. Yet despite these pressures – and sometimes in response to them – science is becoming a highly distributed, international and collaborative endeavour. This kind of scientific collaboration has provided a fertile model for other efforts at We-Think. So will science spawn new and more sophisticated approaches to collaborative creativity that continue to feed the rise of We-Think? There are several reasons to think so.

Scientific eras are defined by the tools scientists have used to gain insights into the world. Newton used calculus to estimate rates of change. Galileo used the telescope to study planetary movements. In the 20th century scientists used X-rays and powerful microscopes. Today, computer power is being applied to science on an unprecedented scale to solve complex scientific problems, and the computerisation of science is promoting more collaboration. Research is being dispersed around the world. Places that were until recently on the margins of scientific research are becoming more central, especially in emerging sciences such as biotechnology. Jeffrey Wandsworth, director of the US Oak Ridge National Laboratory, told a biotechnology conference in Seoul, South Korea, in March 2006,

> The barriers to doing excellent science in the biotechnology, nanotechnology space are falling. Really only about a $300 million investment in infrastructure is needed. This is an area where any country could compete and the possibilities are wide open.

More computers are generating more data that needs to be stored collaboratively. The Sanger Centre at Cambridge, for example, which co-led the Human Genome Project, hosts about 150 terabytes of data – that's about 150 trillion bytes and has processing power of 2.5 teraflops. Sanger is generating data faster than it can analyse it. Floods of data of this kind will be properly organised and fully exploited only if they are open to many researchers. This has already happened in astronomy: until very recently astronomical research was organised around data held by separate institutions; now it depends on shared access to vast data sets held in common. As scientific experimentation increasingly relies on software for simulation, analysis and prediction, so scientists will have to share programs to allow others to test their results. Scientists are pioneering grid computing, so as to yoke together distributed computers to tackle complex problems. One such project is Gloriad, a computer ring connecting researchers in Chicago, Moscow and Beijing as well as offshoots in Amsterdam and Hong Kong. Another is the US Teragrid, which claims to be the world's largest, fastest-distributed infrastructure for scientific research, linking all main centres in the US, using mainly open-source software.

The tools of science are encouraging collaboration – and so is the nature of the problems that science now addresses. Karl Popper, the philosopher of science, distinguished between problems that are like clocks – complicated but soluble – and problems that resemble clouds – diffuse and complex. Science is increasingly about clouds, dealing with questions about how a cell sends messages, how the brain works, how epidemics unfold or ecosystems collapse, how the climate changes and the universe expands. The more science is about complex problems, the more it will require collaboration between disciplines. A prime example is the multinational effort to research the polar regions, which was the brainchild of Karl Weyprecht, a German naval lieutenant who commanded a ship in the Austro-Hungarian Arctic expedition of 1872–74. In the first International Polar Year (IPY), in 1882, 11 nations established 14 research stations to record data on meteorology, terrestrial magnetism and the aurora. The fourth IPY, due to run from 2007 to 2008 – there were two others in 1932 and 1957 – will cost $1.5 billion and involve 228 projects, with 10,000 scientists and 50,000 support staff, in a vast interdisciplinary effort to understand the polar regions, covering geophysics and ecology, social science and economics. This IPY will also be unprecedented its openness: research data will be published as quickly as possible on its website, where users can also download software that will allow them to track the migrations of polar animals.

A further force for collaboration is the way that science publishing is changing. Science is partly a media business: scientific papers, normally published in learned journals after exhaustive peer review, are one of the main outputs of scientific research. In common with most media businesses, scientific publishing is being heavily disrupted by the web. The format has remained largely unchanged since Henry Oldenberg created the first scientific journal at the Royal Society in the 17th century. Scientists want to get their articles published in a high-visibility, high-impact, peer-reviewed journal because that means more exposure, more prestige and in the long run more funding. For scientists the currency, as in computer gaming, is recognition from their peers. Consequently they submit articles without being paid. Readers pay for journals: that is how publishers make their money. The system worked without controversy until scientific-journal prices started to rise – which they have done by more than four times the rate of inflation since 1986. Higher journal prices make research more costly. Scientists even in the developed world report that it is becoming harder, especially at less prestigious institutions, to get access to the research they need. The situation in the developing world is even more worrying. A 2003 study by the World Health Organization found that in the 75 poorest countries in the world, 56 per cent of medical institutions had been unable to access any journals over the previous five years.

Rising journal prices have sparked a revolt – the open-access movement – which is using the web to publish science in new ways. Open access makes research papers freely available online and allows researchers to search, link, download, print, copy, use, distribute, transmit and display an article for a responsible purpose and with proper attribution. Open access is compatible with copyright; it is just that the authors waive most of the traditional rights. Scientists can retain the right to patent discoveries for which they might get paid. Some new online journals are entirely open-access; others make a free online version of an article available only once the physical journal has been published. To get the article more quickly you have to pay.

Open access increases the impact of scientific research by speeding up the cycle and widening its reach: open-access versions of articles tend to be cited by other researchers on average 50 per cent more than the same version in a subscription journal (it is 250 per cent more in some branches of physics). Research spreads more quickly to more researchers who can then test the findings reported in a paper and develop new hypotheses. Digital versions of research papers allow for much more sophisticated tagging, tracking and monitoring

to show which articles are most influential, and where ideas come from and go to.

In 2007, only 15 per cent of the world's publicly funded research was open-access. But research councils around the world – including five of the eight councils in the UK – have started to adopt open-access mandates, which mean that outcomes of publicly funded research have to be made publicly available. More are likely to follow. The US National Institute of Health, the world's largest non-military research-funding agency, has a budget of $28 billion a year – larger than the GDP of 142 countries – and currently a voluntary open-access code that has been largely ineffective: only 4 per cent of institutions have complied. The Institute of Health is expected to mandate open-access publishing for the research it funds, which would affect perhaps 40,000 journal articles a year. In Europe, CERN has embarked on a campaign to make all physics journals open-access. A string of universities, including Southampton, Berlin, Oslo and Stockholm, have pledged to make all their research open-access. Journal publishers are reading the writing on the wall and adopting open-access business models. Venerable bodies such as the American Chemical Society, Cambridge University Press and the *British Medical Journal* have followed in the footsteps of BioMedCentral and the Public Library of Science in creating open-access journals. Scientists are increasingly turning to blogs, wikis and social networking to let people know about their work before it is formally published.

Open access should particularly benefit the developing world, where Western journals are often prohibitively expensive. As China, India, Brazil and South Africa invest more in scientific research it seems likely they will adopt open-access policies. China, the world's second-largest R & D funder, has mandated that all scientific data collected through publicly funded programmes should be open-access. The Gates Foundation has taken a similar step for data on HIV/Aids. One product of open access will be new global repositories of shared scientific knowledge. Archives of open-access research are being created, such as the eScholarship repository at the University of California. The oldest open-access repository of scientific articles, arXiv, which has been going since 1991, has more than 400,000 articles on physics.

Common pools of scientific data provide a platform for collaborative research over many years. One lesson of the Human Genome Project, the collaboration between more than 2,000 scientists from 20 institutes in six countries, which delivered a complete map of the basic ingredients of human DNA in 2003, is that open and collaborative ways of working acquire a powerful dynamic. The Human Genome Project's approach was based on a set of principles drawn up in 1996 at a conference in Bermuda sponsored by the Wellcome Trust. The Bermuda principles – a manifesto for open-source science – were extended and reinforced at a second conference in Fort Lauderdale, Florida, in 2003 organised by the US National Human Genome Research Institute. A growing family of genomic research projects has followed those open-source principles, including projects comparing the human genome with the genome of other animals; The Cancer Genome Atlas; a map to help identify tiny abnormalities in the genome that are linked to diseases; and a structural genomics consortium, financed by the Wellcome Trust, which is posting data on hundreds of proteins linked to diseases such as diabetes, malaria and cancer. These collaborations involve hundreds of scientists, across several countries and many institutions, who will make their data publicly available.

The most ambitious initiative to create a shared and open

repository of scientific knowledge – a blend of the Human Genome Project and Wikipedia – is the Encyclopedia of Life (EOL), which plans to document 1.8 million species of plants, animals and other forms of life. The project, the brainchild of the biologist E. O. Wilson, professor emeritus at Harvard University, has been funded by the MacArthur Foundation to the tune of $20 million. Jonathan Fanton, the foundation's director, speaking at the project's launch, said the encyclopaedia would be a 'global endeavour, a resource of knowledge created by all, maintained by all and with benefit to all'. As of mid-2007, about 1.25 million pages of information had been scanned into a digital database.

Wilson hopes that with information about species being collected in one place, the gaps in our knowledge will become clearer. Fanton predicts the EOL will help researchers see patterns they have missed, provide a benchmark for estimates of species extinction and speed up scientific research. The encyclopaedia could also provide a new model for experts and amateurs combining to create a shared knowledge base. The EOL will not be a conventional encyclopaedia: information from a range of sources will be mashed up. The core will come from trusted sources of high-quality data, provided by the founding institutions: the Field Museum of Natural History at Harvard University, three other leading US institutions and a consortium of the world's 10 largest natural-history libraries, including London's Natural History Museum and the Botanical Gardens at Kew. The EOL will also include user-generated content filtered by a three-step process of quality control. Only material that meets the gold standard will be visible on the site to everyone. The EOL plans to engage citizen scientists with tools developed with Google Earth to allow people to log observations of specimens.

Science in future will not be solely a collaboration among scientists across disciplines and time zones. More sciences will acquire a following of citizen scientists who will work alongside the professionals. Astronomy is a prime example. Like most sciences, astronomy started with amateurs. When Copernicus moved the sun to the centre of the universe he was only a part-time astronomer. Johannes Kepler, who discovered that planets orbit in ellipses, made most of his money from horoscopes. Yet by the 20th century the pendulum had swung decisively in favour of professional astronomers who had access to huge telescopes – like Jodrell Bank in the UK, or the Mount Wilson Observatory near Pasadena, California, where Edwin Hubble determined that the galaxies are being carried away from one another. Professionals probed the outer depths of space; amateurs concentrated on brighter, closer objects they could see with their own puny telescopes.

That all changed with three linked innovations that gave pro-am astronomers cheap and powerful tools. John Dobson, a one-time monk and lifelong stargazer, created an open-source design for a cheap but powerful digital telescope. Dobson remarked,

> To me it's not so much how big your telescope is, or how accurate your optics are, or how beautiful the pictures you can take with it; it's how many people in this vast world less privileged than you have had a chance to see through your telescope and understand this universe.

As Dobson refused to profit from his invention and never patented it, many companies started making telescopes based on his design, combining them with light-sensitive computer chips that could record faint starlight much more clearly than a traditional photograph. Thanks to Dobson's telescope the earth acquired hundreds of thousands of new eyes, recording events in deep space that would have gone unnoticed by the much smaller body of professionals. The Internet vastly amplified this distributed capacity for exploration by helping these pro-ams to work together.[6]

Global research networks have sprung up, linking professionals and amateurs, with shared interests in flare stars, comets and asteroids. Groups of pro-am astronomers have tracked the weather on Jupiter, found craters on Mars and detected echoes from colliding galaxies as accurately as professionals. Amateurs cannot produce new theories of astrophysics and sometimes do not know how to make sense of the data they have collected. An amateur did not write *A Brief History of Time*. Yet in future, aspects of astronomy will depend on dedicated amateurs working in tandem with professionals, motivated by a shared sense of excitement about exploring the universe. More sciences may find themselves having to go down the same path. Biology may be a prime example – and a very troubling one.

The 20th century was dominated by what physicist Freeman Dyson calls 'grey' sciences, which created machines that made us more powerful: the car, the plane, the steel mill, the generator. In this century the focus for funding and research will shift from machines to living, complex systems. Dyson argues it will not take long for the tools of biotechnology to spread from the laboratory into people's homes, giving millions of amateur plant- and animal-breeders new tools to work with. Gardeners will be able to breed their own roses or orchids by spicing gene sequences together. Pet-breeders will be able create their own kinds of dogs and cats. Farmers will be able to make their crops more resistant to local conditions and more productive. As Dyson put it, alarmingly, in a July 2007 article for the *New York Review of Books*, 'Designing genomes will become a personal thing, a new art form as creative as painting or sculpture.'

This is all the more likely if biotechnology becomes a branch of the software industry. Since 1972 genetic engineering has relied on highly specialised techniques that extract and then recombine DNA from cells. It is not a job for amateurs: it takes staff with advanced skills, lots of practical experience, patience, many reagents and a lot of equipment. That could all change, however, with the advent of synthetic biology which allows scientists to create cells by writing 'DNA programmes' to make living organisms. At the moment, biotechnology recombines what is already available; in future, synthetic biology could allow cells to be artificially created. Blueprints for synthetic biology could easily be shared as software is now. Drew Endy, a professor at MIT, is already teaching his students how to build custom-made bacteria by clicking together a set of 'bio-bricks', which are available open-source through his BioBrick Foundation. Rob Carlson, a professor at Washington University and one of the earliest advocates of open-source to biotech, has proposed that Darpa, the US defence research agency, which helped to get the Internet going, should create something similar for biotechnology.

Dyson, for one, believes we will enter a new period of evolution, in which open-source sharing of code will be extended to genes. This may sound far-fetched but it is not far off. The productivity of DNA-sequencing machines has increased more than 20,000-fold in the last 15 years. The costs of sequencing have halved every 13 months. Soon, it will cost less than one US cent to read a base pair of proteins in a gene and only 10 cents to synthesise them. The technology is already migrating out of the labs into society. The equipment for a DNA lab can be bought on eBay and fitted into an average garage. *Make* magazine, the bible for America's home inventors, has already shown its readers how to do what it calls 'backyard biology'. The genomes of dozens of viruses, including Ebola, Marburg and SARS, are publicly available. In 2002 a group of researchers at Stony Brook University in New York state synthesised an infectious polio virus using DNA fragments they bought through mail order. Drew Endy estimates that the Ebola virus could be made for roughly the cost of a new Volkswagen Beetle. In five years' time it might be the cost of an iPod.[7]

Many will be alarmed at the prospect of synthetic biology's putting new power to create and destroy life into the hands of rogue scientists and madcap amateurs. Software programs with bugs can be recalled and rewritten; real-world viruses cannot. Yet synthetic biology also offers the prospect of huge advances: carbon-free fuels made from biomass, cheaper drugs manufactured in the cells of an organism. Open-source biology will pose issues of security, ethics and control far more troubling than those arising from the chat rooms and wikis of the Internet. Whether synthetic biology turns out to be mainly creative or destructive will depend not just on the science, but on our social organisation, on how the knowledge is owned and controlled. My contention is that in the long run, open and collaborative approaches, which involve effective self-regulation and extensive peer review, will be better for good science and our security than either state control or private ownership.

We-Think is already disrupting the culture, media, software and entertainment industries, which are the fastest-growing sectors of the developed economies and account for perhaps 15 per cent of GDP. Key groups will take up the opportunities of the web with greater skill and enthusiasm than others: pro-ams, young teenagers, scientists, avant-garde artists, rock bands. They will drive forward change. But We-Think will not turn everything upside-down, overnight, for everyone.

However, in many sectors of the economy – business services, manufacturing, energy, engineering – We-Think may yet have a more profound impact than we have seen so far, especially on activities like design and research. Many of the basic economic activities we rely on for heat and warmth, food and clothing, and shelter and transport will be affected only indirectly and over a still longer period. Some areas of the economy, such as agricultural production, extractive industries and personal care services, may be disrupted barely at all. You cannot change a wet nappy with a text message. Yet even if the web changes the way we work and consume by, say, 2 per cent a year for 60 years that will add up to a revolution, eventually.

The web's biggest impact will not be in the rich developed economies where it got going. It will be in the developing world, where millions of people are being lifted out of poverty as web technology becomes more easily accessible. For millions in Asia, becoming affluent will be indelibly associated with being networked and connected, largely thanks to the mobile phone. Gershenfeld's Fab Labs may make little sense in the rich world, but they could work wonders in the developing world. Open-access publishing will disproportionately benefit scientists in the poorest countries. Manufacturers in Asia have most to gain from open-source design communities. New models of public education and health may sound far-fetched in societies that already have schools and hospitals, but they make perfect sense in societies with few teachers and doctors. So the developing world will be crucial to the answer to the next and most critical question: will We-Think be good or bad for us?

6
FOR BETTER OR WORSE?

Who could be against more communication and conversation, participation and collaboration, transparency and free speech? To question the spread of the web would be like being against dolphins, green space and trees and things that are self-evidently good. Yet many sensible and thoughtful people, not just Luddites and cultural conservatives, have grave reservations about the impact and implications of the web.

The world is becoming more cacophonous, with many more voices clamouring for attention. Global communications are bringing people closer together but also inflaming jealousies, resentments and misunderstandings, with more people talking past one another rather than to one another. Worse still, many people fear that the web is indiscriminately empowering fundamentalist populist movements, especially in the developing world: any group with a grievance can use the Internet to draw attention to its cause, spread its faith, recruit new adherents and organise its followers. The web's potential to create collaborative, largely self-organising networks has been taken up most skilfully by terrorist groups with a loose authority structure that allows cells to be financed in one country, trained in another, headquartered in a third, and to operate in several others and convey their message globally with the help of little more than a camcorder. Not far behind the terrorists are religious zealots and neo-Nazis, paedophiles and pornographers, gamblers and organised crime syndicates. The British political website that gets the most traffic belongs to the British National Party: racists are not given room to express their views on television so they use the Internet to promote and organise themselves. The web is a boon for any group that needs to be organised but cannot be seen to have an official organisation; tailor made for shadow networks of shadowy people. Security analysts believe that the web plays a major role in contributing to a world that seems to be more bewildering, threatening, disordered and out of control. During the Cold War, the threats of nuclear annihilation were at least managed by states that controlled the weapons of mass destruction. The world may have been threatening but at least it was tidily organised. Now the scales have tipped decisively away from top-down control, and as technology passes into many hands, so it becomes more difficult to prevent it from being used for terror, destruction and crime.

This chapter addresses these worries to explore whether We-Think culture will be good for us. The obvious answer is that it will depend on who you are, whether you are a publisher or a reader, a musician or a listener, an authoritarian ruler or a campaigner for democratic rights, someone living in the rich connected world or someone struggling to feed yourself in a world where the most basic freedoms are not guaranteed. It is obvious that the web will create winners and losers. To go beyond that and assess whether the web will promote ingredients of what might be regarded as a better society is a far more comprehensive and ambitious undertaking and it needs a yardstick. The most commonly used yardsticks – speed and productivity, material and financial wealth – are not the right yardsticks. Instead I offer this assessment of whether the web will promote basic values that are widely, though by no means universally, shared: will We-Think be good for democracy, equality and freedom?

Democracy

Democracy is a prime example of our capacity for collective innovation: our ability to make binding decisions together to produce social innovations, from the abolition of slavery and child labour to the collective provision of education and welfare. Democracy is deliberate collective innovation because it allows us to govern ourselves by combining our ideas and values. So it would seem to be a natural candidate for improvement by the shared intelligence of We-Think, applied not to playing computer games but to the big issues of climate change, peace in the Middle East and caring for the elderly. What evidence is there that the web is improving our capacity to think and act together democratically?

There is a host of reasons for doubting that We-Think will be good for democracy. The fact that more people might have more opportunities to voice their views does not guarantee better debate. It could just mean more squabbling. Even when people engage in political debate on the web they often talk to people they already agree with. Liberal blogs tend to link to other liberal blogs; environmentalists connect with other environmentalists. The web can fracture democratic debate into partisan spaces where people of like mind gather together; democracy depends on creating public spaces where people of different minds debate and resolve their differences.

Politics is increasingly a highly sophisticated, professional, full-time occupation. Only an equally vigilant, professional cadre of journalists and independent regulators will keep politicians in check. Yet the web could erode support for this professional scrutiny. As more people turn to amateur blogs and online news and away from newspapers and mainstream television news, so funding for professional, investigative journalism could decline. Instead of facing hardened, diligent journalists who know how to dig away at a scandal, politicians could easily lord it over a Lilliputian rabble of ill-equipped, amateur bloggers who can be easily ignored. That would weaken democratic scrutiny.

As more mainstream politicians take to the web, with their carefully calculated YouTube channels and social-network profiles, so they could diminish its radical potential. The web could become a tool for politics as usual. And even if the web does not benefit the old élite it could well create a new élite to take its place, the technorati who are adept at using the web for political purposes. Newspaper commentators may have less influence; superstar bloggers may have more: ordinary citizens may be no better off. And even if élites are not exploiting the web to entrench their power it could be taken up by populist movements that have little interest in deepening democratic debate and may instead promote a form of digital mob rule. One thing is for sure, digital technology alone does not make a society more democratic. Japan is one of the world's most technologically enabled societies, where the high-speed mobile Internet is ubiquitous. Yet Japan's pork-barrel politics have been changed little by technology. China is fast becoming the world's largest Internet-connected society, but as of 2007 the Communist Party was still keeping a tight grip on its use for anything vaguely resembling political purposes. At the most the web is just one small influence on political culture. Yet we should not discount the web's democratic potential just yet.

In the developed world, formal party politics is an industry in decline. In the late 19th century, all over the industrialising world people campaigned to be allowed into political processes that were dominated by élites. They demanded and slowly were given the vote. Little more than a century later, the disenchanted great-grandchildren of those campaigners are flooding out of the political realm as fast as they can. In 1960, 62.8 per cent of Americans of voting age went to the polls; by 1996 it was 48.9 per cent. Membership of US political parties halved between 1967 and 1987; politics was fast becoming the pursuit of a tiny minority. In Britain four out of 10 people chose not to vote in the 2001 and 2005 general elections, rising to six out of 10 among 18–25-year-olds. The 1997 election attracted the lowest turnout since 1945. By 2007 fewer than 2 per cent of the population were members of the main political parties and membership was less than a quarter of its 1964 level. A more individualistic, consumerist culture has eroded the collective identities that mass political parties were based on. Government seems more distant from the intimacy of people's lives and yet less able to protect people from global forces. People talk of their political representatives as invisible, alien, partisan, arrogant, untrustworthy, irrelevant and disconnected.

That decline in political engagement has coincided with the spread of mass media and broadcast television, which provide the information backbone to our public life and are where issues are debated and politicians appeal for our attention and support. Newspapers and television have high fixed costs – print plants and television studios – so they have to reach mass markets to earn their keep. As a consequence, industrial-era media suffer from serious weaknesses as a vehicle for democratic life.

Democracy depends on popular sovereignty, which in turn depends on free speech and open discussion among citizens.

Anything that restricts speech – the limited number of pages in a newspaper, restricted television airtime, a narrow radio spectrum – creates inequalities between those whose views are aired and those whose are not, and impedes the free flow of information, so undermining the quality of democratic decision-making. A small group of news editors decides which views and voices are to be heard, when a story is important and when it drops off the bulletins. During the two months of the 2000 US presidential election the three major network news channels – ABC, CBS and NBC – devoted 805 minutes to campaign coverage, an average of four minutes per night per news programme. Robert Putnam, the social capital theorist, blames television's couch-potato culture for much of the decline in civic participation in the US. Television turns people into an audience to be titillated and entertained rather than citizens who have a responsibility to engage in debate. Television rewards politicians with style and emotion over substance and ideas. It also heavily skews democratic politics in favour of the rich. Industrial-era mass media concentrate ownership and give undue weight to the views of a few proprietors who own the printing presses. Television airtime is costly. To pay for television advertising candidates have to raise large sums of money, which makes them reliant on rich donors and opens up opportunities for corruption as policies are tailored to their interests. It is not surprising that so many people are turning away from a political process that offers opportunities of influence to so few. That is the context in which the political impact of the web needs to be judged. Can it revive a democratic politics that at least in the US and the UK has become so hollowed out?

The web's most optimistic political advocates believe it could be an elixir for democratic life, making relations between politicians and voters more direct, debates more deliberative and citizens more engaged. As long ago as 1984, Benjamin Barber in *Strong Democracy* forecast that future technologies could be used to 'strengthen civic education, guarantee equal access to information, tie individual and institutions into networks that will make real participatory discussion and debate possible across great distances'. As the Internet developed in the 1990s, apostles such as Dick Morris, President Bill Clinton's 1996 campaign manager, proclaimed that the web would save democracy by making it more direct.[1] Voters would comment and vote on proposals as they passed through a virtual legislature, cutting out all parties and the mainstream media.

The other main political claim for the web is that it should make democracy more thoughtful and deliberative, a line of argument derived from the work of Jürgen Habermas, the German political philosopher. Habermas argues that free and undistorted communication should provide the basis for a true democracy based on perpetual conversation between citizens in which anyone can join, can raise a topic for debate or question the rules for conducting the discussion. The web's best known self-governing communities do seem to bear this out: they rely on responsible self-governance, in which decision-making is relatively transparent and power held to account. The outcome should be that citizens will become more engaged and the political process more legitimate and creative. Yochai Benkler, a Harvard law professor, argues that as the web allows more views to be aired, so citizens will become more critically engaged in debating and choosing between a wider range of proposals. Benkler's approach echoes Hannah Arendt's view that citizens should be the 'craftsmen of democracy', inquiring into how things work, getting their hands dirty, challenging

the power of experts and professionals who dominate policy-making. Arendt advocated a kind of what we might now call Homebrew democracy in which everyone had the right to think aloud, debate with and challenge others regardless of whether they were an expert or an amateur.

So, according to its proponents, the web should revitalise exhausted party politics by shifting information and power from élites – whether professional politicians, parties, commentators or policy wonks – to those formerly consigned to being spectators. That should allow a more diverse range of people to participate more fully in democratic debate and create ways to campaign, debate, deliberate and scrutinise which will prove more productive, legitimate and creative. Is that what is happening?

It is still too early to decide whether the sceptics or the optimists will be proven right. Radio became a political tool when Franklin D. Roosevelt started using it for his fireside chats in 1933. Television came of age when the advertising dream image of John F. Kennedy defeated the camera-challenged Richard Nixon during live debates in 1960. The web entered mainstream politics only in the mid-1990s: the first politician to create a website was Diane Feinstein, the Democratic vice-presidential candidate, in 1994, the same year Anne Campbell became the first British MP to create one. In the UK the main political parties had websites in the 1997 general election, and a year later 60 MPs had their own. By 2002 about 20 per cent of MPs' correspondence was arriving by email and in 2003 three MPs had blogs and 260 had websites. In 2006 both main party leaders had channels on YouTube, 560 MPs had websites and 39 were blogging. However, it took an outstanding political failure – Howard Dean's – to highlight the web's potential to breathe new life into politics.

Dean was an outsider in the race for the 2004 Democratic presidential nomination and when he set off, he lacked an organisation. The only way Dean could make an impact was to turn his supporters into participants and contributors. Dean's campaign manager, Joe Trippi, and a team of young volunteers had no option but to create an Internet-based campaign that relied on local self-organising support groups and a mass of small donations. This was a political campaign that followed the pattern of a multi-user computer game, with the central campaign team playing the role of Electronic Arts and the supporters becoming player-developers, enriching the game, extending its life, adding new energy. By the time Dean met his demise the campaign had mobilised $60 million in donations and perhaps 500,000 political player-developers.

The Dean campaign had many of the traits of We-Think. It was easy for people to join in and pick up tools to allow them to get working without asking for central direction or even permission. The campaign used social-networking software that allowed contributors to form local chapters. People were encouraged to make any kind of contribution they could, no matter how small. The campaign then clicked these mini-contributions together as if they were Lego bricks. After one supporter emailed that she had sold her bike to raise funds for Dean, thousands of others followed suit. Others formed Dean Corps, who cleaned up rivers and collected food for the homeless. In his memoir of the campaign, Trippi argued:

> Power is shifting from institutions that have always been run top-down, hoarding information at the top, telling us how to run our lives, to a new paradigm of power that is democratically distributed and shared by us all.[2]

Trippi imagined that a much more engaged democracy would emerge from campaigns like Dean's. People will not wait to be fed information; they will go and find it themselves. They will not be content to be told what happened; they will question official accounts. They will learn from one another as much as from figures of authority and they will expect leaders to take part in their conversations and to use their authentic voice when they do so. Above all they do not want to be talked down to or lied to.

Trippi hoped that the Dean campaign would usher in a new kind of politics based on a dialogue with and among the voters, drawing on the collective intelligence of thousands of supporters:

> It became obvious pretty quickly that a couple of dozen sleep-deprived political junkies in our corner offices could not possibly match the brainpower and resourcefulness of six hundred thousand Americans.

What evidence is there that the web is living up to these hopes and breathing new life into an otherwise exhausted political domain?

The web does seem to be drawing a wider range of people into democratic debate, many of whom might not otherwise have become involved in politics. Between November 2002 and November 2006 the share of the US adult population with broadband connections at home rose from 17 per cent to 45 per cent. Those who are broadband connected are using the Internet to engage in more political activity than they did before. The proportion of Americans who turn to the Internet as their primary source of news rose from 3 per cent in 1996 to 15 per cent in 2006. That goes up markedly when there are elections: during the 2006 congressional election campaign, 46 per cent of regular American Internet users and 31 per cent of the population as a whole gathered information and exchanged emails about politics. The make-up of the online political class has changed. In 1996 it was mainly geeks: college-educated, white, young, male, suburban and middle-class. In 2006 it was still mainly college-educated and middle-class, but in many other respects – gender, ethnicity and geography – it mirrored the population as a whole.

The Internet is especially effective in drawing young people into politics. Almost all the young people who donated to political causes during the 2004 US presidential election cycle did so online. According to the Pew Internet & American Life study of the 2006 congressional elections, 11 per cent of Internet users and 5 per cent of the population as a whole – about 14 million people – regularly used Web 2.0 tools to engage in political debate and campaigns. Online donors tended to stay engaged: the study found they were much more likely to donate again than those who donated offline. Bloggers were much more likely to sign petitions, attend rallies or write to their Congressmen. The online political class is influential among friends and peers because it is highly networked offline as well, according to the Pew study. It is not all seasoned political activists taking to the web, though. Of the 400,000–500,000 people who attended political meetings organised via Meetup in the 2004 election, half had never been to a political meeting before and 60 per cent were below the age of 40. A study of those elections by the Institute for Politics, Democracy and the Internet found that 44 per cent of Internet users who were politically engaged had not been so in the past. The 2008 campaign is likely to show a further extension of these trends.

The Internet is drawing more people, especially young people, into political activity – albeit still a minority of the population. Is that changing who influences the political agenda, for example by reducing the power of big corporate donors? In the US, the candidates with the most to spend on advertising tend to win the nominations in presidential and congressional races. In the 2004 primary season, George W. Bush and John Kerry each raised about $500 million for their campaigns. Yet this money comes from a tiny share of the US population: in the 2004 campaign only 0.52 per cent of the population of 300 millon made contributions of more than $200, and 0.12 per cent made contributions of more than $2,000. The funding system gives a small minority of the electorate a disproportionate influence. If the Internet can change how politics is funded then it will change something very significant.

Howard Dean was not the first political candidate to use the Internet to raise considerable sums in campaign-funding – Bill Bradley, John McCain and Minnesota governor Jesse Ventura did so as well. Yet Dean's bottom-up campaign showed the potential for gathering small donations to match the clout of large donors. Dean's only hope was to find a way for ordinary people to make small contributions that could become significant when added together: not so much We-Think as We-Fund. In one day the Dean campaign managed to raise $500,000 in small online donations from 9,700 people, while Dick Cheney was raising $250,000 having lunch with 125 bigwigs. Ultimately Dean raised more than $30 million in donations of about $50 each.

In the 2008 campaigns for the presidential nominations, Dean's pioneering techniques became standard practice. In the first six months of 2007 the three leaders in the race for the Democratic nomination – Clinton, Edwards and Obama – raised $28 million online, mainly in the form of small donations. Obama raised more than $17 million online, 90 per cent of which came from 110,000 contributors each giving less than $50. By mid-2007 Obama had more than 9,000 'micro-bundlers', each with a page on his campaign website, on which they put the email addresses of friends and contacts, who could then be approached for donations. Small online donations are certainly not revolutionising US politics. The candidates in the 2008 elections needed more money than ever, and the lion's share continued to come from rich donors. But the Internet is allowing non-mainstream candidates to play more of a role in the campaign and giving ordinary people a new way to connect to politics. A candidate's ability to raise money from small donations is becoming an important sign of his or her political reach and legitimacy. A candidate who cannot raise money this way would look suspiciously in hock to large donors.

Is politics becoming more open and responsive to a wider range of views, issues and voices thanks to the web? US politicians are much more aware of the political commentary provided by blogs than they were: a 2006 survey found that 90 per cent of congressional staff were aware of or read blogs. Yet bloggers are most effective in conjunction with the mainstream press rather than on their own. A study of the 2004 presidential elections found that bloggers generally followed a story rather than creating it. They have been most effective acting as a corrective to the shortcomings of mainstream media. The earliest publicised instance of bloggers' combined ability to act as a public watchdog was the resignation of then Senate majority leader Trent Lott after he had made racist remarks at the 100th-birthday party of hard-right Republican

Senator Strom Thurmond. The remarks, initially ignored by the mainstream media, were taken up by liberal and conservative bloggers who forced the story into the mainstream media. Since then more politicians have fallen foul of the bloggers' ability to pick up virtually any indiscretion.

In the 2008 US election YouTube provided another powerful tool for citizen journalists and commentators, propelling people like James Kotecki, a recent graduate who goes under the name EmergencyCheese, to Internet fame. Kotecki's acerbic video commentaries on the candidates' efforts to connect with younger voters through YouTube started to elicit responses from the candidates themselves. YouTube seems to be shifting the balance of power, however slightly, from politicians to people. Political videos created by YouTube users have been far more popular than anything provided by the candidates. One infamous posting based on an old Apple advertisement, which compared Hillary Clinton to Big Brother from Orwell's *1984*, was viewed 3.5 million times between March and July 2007 and attracted more than 10,000 comments. The Clinton campaign's own offerings fell well short of this reach. Clinton generated a huge amount of press coverage by asking her supporters to choose a campaign song via a spoof of the final episode of *The Sopranos* in which she appeared with her husband. The most-watched version of this on YouTube had by July 2007 been viewed 276,000 times, and elicited 751 comments.

Phil de Vellis, the creator of the *1984* spoof, told the *New York Times* that he made the advertisement because he was

> frustrated by the way politicians are using video, treating it just like they treat TV and I wanted to make a statement saying that you have to do more, you have to actually interact with your audience and a pretend conversation is not enough.

David Counts, creator of the 2004 website www.toostupidtobepresident.com, explained his motivation as being, similarly, 'to say something amid the smothering, uniform, corporate media messages'. Thanks to blogging and increasingly to YouTube some citizens can play more of a political role, holding politicians to account, knocking them off their pedestals and forcing them to listen. As a result the web may also be making politics more plural, allowing for more niche positions.

Television forces politicians to appeal to very large audiences, and that drives them to sound alike, to cluster in the centre ground. Online politics allows politicians to engage with much smaller, committed communities and tailor policies to them. That is one reason why, by mid-2007, Hillary Clinton had a growing list of commitments on dairy farming, nutrition in children's cereal and bias in school textbooks. In future, politicians may have a handful of policies that appeal to the political mass market and a long tail of policies appealing to smaller communities. As more people participate online, creating profiles on Internet social networks, they leave trails of personal information enabling the creation of powerful databases that politicians can use to target particular groups. In 2004 George W. Bush did surprisingly well among black male voters in Ohio because micro-targeting techniques predicted they would respond to messages about education and health.

Ultra-local politics may also thrive with the web. Keith Hampton, a professor of sociology at the University of Pennsylvania, who has studied neighbourhood social networks in Toronto and Boston, found that people involved in local online networks were more likely to get involved in local politics, from tackling crime and social disorder to making complaints about public services. Hampton believes social networking could help to reverse the decline in social capital that Robert Putnam blames for civic disengagement. Hundreds of local discussion groups and social networks have formed in the US in last five years. In Chicago, a city council project called Full Circle allows local communities to create their own maps to help them plan their own strategies for renewal. One Latino neighbourhood, for example, used the tool to map local restaurants and food stores, and found that all but three of the restaurants served fried food and only two of the shops carried fresh fruit and vegetables. In the wake of Hurricane Katrina the Internet group America Speaks organised a series of on- and offline conferences among displaced residents of New Orleans, to debate plans for the city. Ultra-local political discussion may be a significant beneficiary of the web.

What the web is not doing, however, is creating a more deliberative, conversational democracy at a national level, in the US or elsewhere. In Estonia, the European state with the most Internet-enabled political system – all cabinet papers are available online and cabinet decisions are reported on the web as they happen – officials say this unprecedented degree of transparency has chiefly speeded up the political process rather than making it more thoughtful. Yet even if the web does not prove to be a platform for deliberative democracy it is already proving to be a powerful way for people to mobilise in campaigns: not so much We-Think as We-Act. The web makes it much easier for people to connect with each other to sign petitions, attend rallies and donate money to causes they believe in. After all, democratic advances rarely come from reasoned argument alone; invariably they require protest and campaigning to force those in power to change tack. That is why the web could be good for democracy: it opens up the field to a new, wider range of campaigners.

Howard Dean's campaign highlighted this potential through its alliance with the website Meetup which makes it easy for people of like mind to meet, usually in groups small enough to fit into a Starbucks. When the Dean campaign became aware of Meetup it had 432 supporters from across the US wanting to meet to talk about Dean's policies. Eventually it had 190,000 members registered on Meetup, and about 500,000 people attended Meetup organised rallies. Most of the 2008 candidates followed in Dean's footsteps. Ron Paul, the Republican outsider, had 552 Meetup groups by July 2007, with 24,000 members who had held 1,400 events.

To see the web's power to mobilise people in political action in their millions, especially when that power is combined with the potential of mobile phones, one needs to look outside the US. Mobile phones are spreading around the world more quickly than any previous technology, especially in Asia, but also among the poorer and younger parts of the population of developed economies, including recent immigrants. The Internet-connected mobile phone may do more than the computer to draw people into politics. In 1991, there were 34 land-line telephones in the world for every mobile phone. By 2004, mobile-phone subscriptions had overtaken land-line subscriptions for the first time: 1,748 million mobile lines as against 1,198 million land-lines. In January 2004, 52 million Africans had mobile phones, while just 5.8 million used email. The potential growth in mobile-phone penetration is huge, especially in the developing world.

While the US spawns innovation in the computer-based web, it is developing economies, such as the Philippines, that are most innovative in the use of mobile phones. The average Filipino mobile-phone user sends over 2,000 text messages a year. The country's 30 million mobile-phone users send 200 million messages *a day*. As of late 2004 only 27 million US mobile-phone subscribers used text regularly. The rapid spread of mobile communications in the developing world is creating new political possibilities: independent channels of autonomous communication person-to-person that are difficult for the authorities to jam. A technology that can mobilise friendship networks for political ends is potentially very powerful.

In the Philippines in January 2001, four days of popular protests in Manila, involving thousands of mobile-phone-touting demonstrators, led to the removal of President Estrada who was facing corruption charges. Debate and gossip about Estrada's corruption had accumulated from 1998 in online forums and chat rooms. By 2001 there were 200 websites devoted to the subject and more than 100 email discussion groups. One, E-Lagda.com, collected a petition with 91,000 signatures demanding Estrada's resignation. The result of those protests was that Gloria Arrayo, a Harvard-trained economist, was sworn into office. Arrayo was hounded from office in 2004, in part thanks to a 17-second mobile-phone ringtone that was a recording of her arranging to rig the forthcoming election. The ringtone was downloaded 1 million times from the website of Txtpower.org, which has become a political force in its own right in the Philippines.

Other examples of the power of 'mobile web politics' include the December 2002 elections in South Korea, which were won by President Roh Moo-Hyun largely thanks to his online supporters' group Nosamo. On the day of the election, 800,000 emails were sent to mobile phones urging people to vote – and this swung the result in Roh's direction.

Mobile politics also had a decisive impact on the Spanish elections of March 2004 after radical Islamist terrorists bombed three suburban trains in Madrid, killing 192 people. Soon after the bombings, which occurred on Thursday 11 March, the governing party, the Partido Populaire (PP), blamed ETA, the Basque terrorist group, an accusation widely reported without question by the mainstream media. The day afterwards, the government organised demonstrations of solidarity against the attacks but, by that stage, allegations that the government was manipulating the bombings for its own ends had started to surface on the web. On Saturday 13 March, a text message began to circulate urging people to congregate outside the PP offices in Madrid for a silent protest. Text-message traffic in Spain was 20 per cent up on an average Saturday; on the Sunday it was 40 per cent higher than on an average Sunday. The protests spread from Madrid to Barcelona and eventually to every major Spanish city. On 14 March, the PP lost the election.

In the Ukraine, Georgia and Kyrgyzstan, 'colour revolutions' created by popular movements networked by mobile phone have overturned unpopular or fraudulent election results.

Campaigns that mix mobile phones and social networks are starting to spread to the US. In 2006 a march in Washington against proposed anti-immigration legislation prompted a string of protests in other cities. As one report from Houston put it,

> They can get a crowd to assemble someplace within half an hour, a crowd of tens of thousands of people, simply by everybody text messaging five people.

One of the organisers of the California protests remarked,

> I think MySpace and cell phones were 95 per cent responsible for organising the protests all over the state.

And a newspaper reported from Las Vegas,

> Police and school officials said at least 3,000 students, drawn together by text messages and cell-phone calls, left high schools, middle schools and community college after the morning bell to join the protests.

Just as industrialisation and urbanisation created new political players, the trade unions, friendly societies, co-operatives and ultimately the labour movement, the web will create its own political actors: networked campaigns that might be short-lived and focused or more enduring. One early example which has inspired imitators in Australia and the UK is MoveOn, the US network born when two Internet entrepreneurs circulated a petition against Republican efforts to impeach President Clinton, calling on Congress to instead 'move on' to address other more pressing issues. Within a week 100,000 people had signed up. Co-founder Joan Blades recalled,

> We thought it was going to be a flash campaign, that we would help everyone to connect with leadership in all the ways we could figure out and then get back to our regular lives. A half a million people ultimately signed and we somehow never got back.

As of early 2007 MoveOn had more than 3.3 million members across the US, with more than 268,000 active volunteers and 700,000 individual donors – and only 15 staff. The movement is involved in an eclectic mix of issues from campaign-finance reform to environmental protection and social security. It helped to block efforts to remove federal funding from National Public Radio (NPR) and the Public Broadcasting Service. In the

2006 congressional elections, MoveOn volunteers, working from home and co-ordinated by email, made 7 million phone calls, hosted 7,500 house parties and ran 6,000 events in target districts in support of progressive candidates. MoveOn is neither a single-issue campaign nor a political party, neither a pressure group nor a flash mob. It is a new kind of political animal spawned by the web. There will be more.

In the next 10 years, the web will open up a huge opportunity for people to create a new generation of civic organisations, often at low cost. The Internet will create new political players. As Pippa Norris, a political scientist at Harvard, argues,

> The many-to-many and one-to-many characteristics of the Internet multiply manifold the access points for publicity and information in the political system. The global dimension of the Web facilitates transnational movements transcending the boundaries of the nation state. The linkage capacity strength ens alliances and coalitions.[3]

Finally, but most significantly, the web is likely to spread democracy into repressive and authoritarian regimes, especially in Asia, the Middle East, North Africa and the former Soviet Union. Even if it does little to revive the flagging democracies of the West, democracy may gain a foothold in authoritarian states. Internet penetration in many of these societies is very low. Less than 1 per cent of the population of the Yemen have access to the Internet; at the end of 2006 there were fewer than 300,000 computers in the whole country. Yet access is spreading fast. In 2007 Africa recorded the fastest growth in broadband connections worldwide, albeit from a low base. The proportion of the Vietnamese population with Internet access grew from 5 per cent in November 2004 to 17 per cent, or 13 million people, just two years later. In many developing world societies, people connect online through cyber cafés, so the number of computers connected to the Internet vastly underestimates usage. China will become the world's largest Internet society sometime in the next decade: by 2007 it had 130 million Internet users, compared with 190 million in the US. Only 4 per cent of India's 1 billion people are connected, but within a decade it could have trebled.

Governments in developing countries are keen to promote the Internet to connect their societies to global flows of technology and business. But many are also alarmed by the freedom connection gives citizens to conduct their own debates and question government policy. So across the Middle East, much of Asia and the former Soviet Union governments are clamping down on the use of the Internet for political purposes in the name of national security. A 2007 report by the Open Net Initiative, which brings together researchers from Toronto, Harvard and Oxford universities, found that governments in 25 countries, including Azerbaijan, Burma, Iran, Saudi Arabia and much of the rest of the Gulf, filter the Internet to prevent access to political material.

China runs the most sophisticated censorship regime, mainly using technologies provided by Western companies such as Cisco. Every website in China has to be registered with the authorities. Cyber-café owners are enlisted as spies. There might be as many as 54,000 Chinese police working in cyberspace. Internet service providers have to comply with state regulations on content. Access to websites using words like 'democracy' and 'freedom' is restricted. One victim of these restrictions was Shi Tao, a journalist, who used his Yahoo account to email a pro-democracy website in the US detailing government restrictions on newspapers reporting on the anniversary of the 1989 Tiananmen Square demonstrations. Shi Tao's email led to his imprisonment for 10 years' hard labour. Iran has blocked access to Wikipedia and the New York Times website and announced in late 2006 that it would monitor all citizens' web usage. In Burma the government makes broadband connections prohibitively expensive and slow; dial-up Internet access comes with a limited menu of state-approved websites and email is charged per message. Despite these restrictions, the pro-democracy demonstrations of autumn 2007 attracted attention from all over the world mainly thanks to pictures, video and reports compiled by bloggers giving a minute-by-minute account of events.

The restrictions imposed by authoritarian regimes are a testimony to how much they fear the Internet. In such countries the web will provide the main space in which democratic dissidents will gather. In Vietnam, for example, the opposition People's Democratic Party was founded on the Internet in 2005 and is organised online, partly using voice-over Internet phones. Bloc 8406, a dissident group, launched an online petition in April 2006 signed by 118 democracy activists; it has since been signed by thousands more. Without the Internet there would be little or no democratic opposition in places like Vietnam.

The best measure of the web's political significance is not how many friends Barack Obama has on Facebook; it is whether bloggers in Syria and Burma, China and Iran can raise their voices. That is why it is so vital to preserve the Internet as an open global commons for the exchange of information and ideas. Or look at it this way: by mid-2007 the US government had spent more than $350 billion on the war it started in March 2003 to bring democracy to Iraq and then to the rest of the Middle East. Yet only 4 per cent of people in the Arab world have broadband access. The most potent way to promote democracy in the Middle East would be to get that figure above 50 per cent. The biggest contribution the Internet could make to democracy would be to propel an orderly transition away from one-party rule in China.

In December 2005, news filtered out through the Internet that Chinese police in Shanwei, Guangdong province, had killed several villagers who had been protesting against a wind farm that threatened their livelihoods. To suppress news of the incident the authorities had to shut down cyber cafés in neighbouring areas, cut off Internet access to residents, impede queries for the town's name on search engines and erase blog mentions of the incident as soon as they appeared. Despite all that, a human-rights group investigated the incident and posted its report online. The battle to bring democracy to repressive states will be fought online through thousands of struggles like those in Shanwei, by people who want simply to be able to think out loud, together.

So on balance, will the open web be good for democracy? Yes, it will.

Equality

The Utopian hopes invested in the web's democratic potential are matched by claims for its capacity to promote equality by breaking down concentrations of power based on information and knowledge, and lowering barriers to entry into the market-place for ideas. In a global economy that trades information and ideas as much as raw materials and physical goods, anyone with a computer and a modem can become a participant. At least that is the theory.

There are serious doubts as to whether the web will do much

to make the world less unequal and make any difference to the most pressing problems facing the poorest societies in the developing world. About 40 per cent of the world's population lack access to basic sanitation facilities; more than 1 billion people cannot get safe drinking water; 80 million children – the majority of them girls – are not receiving even a primary education; and 95 per cent of the 40 million people worldwide living with HIV and Aids are in developing countries. What difference would it make to these people to get a profile on MySpace? You cannot feed a hungry child with MP3 files. Even where the web might play a role by providing access to information and knowledge, it will reflect inequalities rather than lessening them. Servers, routers, cable, hard drives cannot be wished out of thin air. They have to be bought. The people most able to take up the opportunities the web creates are the ones with the financial resources to do so. In high-income countries in 2003 there were 430 computers per 1,000 population, but in the poorest countries only 6 per 1,000. High-income countries account for 15 per cent of the world's population but 95 per cent of Internet connections. Worse still, the collaborative nature of the web might reinforce these inequalities by helping the well-connected to become even-better-connected. An extensive study of social networks on Facebook by researchers at Michigan University, for example, found that those with strong social networks use the web to strengthen them further. Those with knowledge and power can network even more with one another. In short, according to the sceptics, the web is at best irrelevant to inequality; at worst it reinforces it; and mainly it is merely a mirror for how unequal society is.

Assessing whether the web is good or bad for equality is fraught with difficulties. Are we interested in equality of access, opportunity, outcomes or capability? Equality between whom – men and women, young and old, rich and poor, within societies or between them?

To make the task more manageable this section adopts a simplified version of the formula made famous by the political theorist John Rawls in his book *A Theory of Justice*: social justice is advanced only if an improvement in the living conditions of the richest benefits the poorest even more. What that means is that the rise of We-Think culture among affluent teenagers in California might be justified on grounds of social justice if provision of the same tools could make an even bigger difference to the lives of the poorest. Then we could say the spread of We-Think potentially makes the world more socially just.

There are grounds for guarded optimism, in theory. As Yochai Benkler argues in *The Wealth of Networks*,

> Information, knowledge and culture are core inputs into human welfare. Agricultural knowledge and biological innovation are central to food security. Medical innovation and access to its fruits are central to living a long and healthy life. Literacy and education are central to individual growth, to democratic self-governance, and to economic capabilities. Economic growth itself is crucially dependent upon innovation and information. For all these reasons information policy has become a critical element of development policy and the question of how societies attain and distribute human welfare and well-being. Access to knowledge has become central to human development.

We-Think could be good for social justice in several ways. As Nobel prize-winning economist Amartya Sen argues in *Development as Freedom*, democratic governments are less able than dictatorships to cream off resources for an élite and ignore the plight of the poorest. Formal democracy, Sen argues, relies on a wider culture of dialogue and argument:

> Discussions and arguments are critically important for democracy and public reasoning ... the argumentative tradition, if used with deliberation and commitment, can also be extremely important in resisting social inequalities and in removing poverty and deprivation. Voice is a crucial component of the pursuit of social justice ... The critical voice is the traditional ally of the aggrieved and participation in arguments is a general opportunity not a particularly specialised skill.[4]

If the Internet is good for democracy in the developing world, it should then also be good as a check on the abuse of power that entrenches inequality.

Open and collaborative approaches to science and innovation have the potential to benefit the developing world more than traditional proprietary and commercial approaches. If we could find a way to spread knowledge and ideas more successfully at lower cost then more people would benefit. The real significance of Wikipedia, largely overlooked by its critics, is that it is creating a global source of knowledge, in scores of languages, that everyone with an Internet connection can access for free. We may be developing a new way to provide public knowledge on a global basis at very low cost. An information commons of that kind ought to be good for development in the poorest regions of the world.

Commercial approaches to research and innovation suffer from two drawbacks as far as equity is concerned. Proprietary systems for owning and controlling knowledge limit direct research to topics of interest or concern to those who can pay. That is why so much pharmaceutical research is devoted to diseases of the rich and corpulent and so little to diseases of the poor that affect many millions more people. In most science, one person's research output becomes another person's inspiration or input. If proprietary controls – such as patents and copyrights – make accessing ideas more expensive, such as by putting up journal prices, they will price out of the market innovators who cannot afford to pay to access the knowledge. Most developing countries spend less than 0.5 per cent of GDP on research and development. In some African nations, the people with postgraduate science degrees can be counted on the fingers of one hand. The highest-income countries, members of the OECD, have 3,281 scientists and engineers per million of population, compared with 788 per million in developing countries, according to the World Bank *World Development Report* for 2003. Patents granted to residents of high-income countries average about 346 per million of population, but in the poorest countries it is about 10 per million. Collaborative and open-source production information and knowledge could help redress these imbalances.

An experiment in open-source agricultural biotechnology, helping to feed people more efficiently, is a prime example of what is becoming possible. When a plant emerges from the soil with only its crown showing it can sometimes grow a bulge. Plant scientists initially thought these bulges were caused by a bacterium that introduced a toxin into the plant. In the 1990s, geneticists discovered that the bulges were actually tumours, caused by a gene inserted by the bacterium which made cancer cells grow. It did not take long for scientists to realise that they could use the bacterium to insert different genes into plants, ones that would make them grow larger more quickly, making agriculture more productive. Within a decade, the large biotech companies had surrounded the process with a thicket of more than 200 patents, which meant that plant scientists in developing countries had to pay them to use the technique. In late 2004, however, after three years' analysis, Cambia, a tiny, not-for-profit organisation in Australia, found an 'open-source' alternative. Cambia identified three other bacteria that did the job just as well and released details of the 'source code' for the technique on the Internet. The aim was to make available to scientists, plant-breeders and farmers across the developing world techniques to genetically engineer healthier maize and disease-resistant crops.

Cambia's founder, the maverick scientist and social entrepreneur Professor Richard Jefferson, said his goal was to democratise innovation in favour of the poor:

> What people do not like about GM is that the technology is controlled by multinational companies and no one outside science understands it and so they don't trust it. We want to put the technology and the science out in the open, in the hands of poor farmers and plant scientists in the developing world who have an interest in using it responsibly and productively. I am a tool developer. I want to develop tools that allow more people to do more science more effectively. Inventing a method that many people can use is far more powerful than inventing a product that people can use. Tools are empowering in a way that products are not.

Cambia has posed a challenge to the way that innovation is traditionally effected in the pharmaceuticals and biotech industry. From his modest base, struggling for funding, always trying to do a thousand things at the same time, his hair going in several directions at once, Richard Jefferson is trying to create a new kind of global public good: open-source

OPEN
SOURCE

biotechnology tools. As he put it when we met at a conference in Brazil in 2005,

> The Internet is revolutionising how knowledge can be shared and how it can coalesce in shared projects. It's not just about sharing knowledge but about creating new ideas through collaboration. Creativity comes from individual flashes of brilliance that then attract hundreds of other people to make contributions. We know that innovations that involve users and developers from the outset are far more likely to succeed than those that do not.

Many other approaches to sharing knowledge for medical development are emerging. Chinese research is leading the way, making available genetic data on rice, cotton and basic crops vital for subsistence farming (as well as spawning a growing biotech private sector). The Public Intellectual Property Research for Agriculture initiative, launched by a group of 14 US universities, pools their patents and makes their research freely available to the developing world. A new generation of public-private partnerships is taking shape in medicine, bringing together social entrepreneurs, scientists and foundations to develop medicines for the poor that otherwise would have been left on the shelf. One of the pioneers of this approach is Victoria Hale, a bright-eyed American drug-developer who jumped ship from mainstream pharmaceuticals to create the Institute for One World Health. Hale's aim is to get pharmaceutical companies, universities and research labs to donate patents and discoveries that they have no intention of commercialising, but which might have an application in the developing world, to tackle afflictions such as TB, malaria or diarrhoea. The Institute's first drug, an injection to cure visceral leishmaniasis – so-called black fever, carried by sandflies, which affects about 500,000 Indians a year – was approved for use in India in 2005. The 'black fever' market was too small for pharmaceutical companies to profit from, even though they had drugs that could do the trick. One World Health developed a collaborative solution with the World Health Organization, the government of India and Gland Pharma, an Indian manufacturer that will make the medicine available at cost – about $10 for a lifetime cure, a fraction of the price charged for other remedies.

The collaborative web does not just lower the cost of research and innovation, it allows people to participate in adapting ideas, and through their participation they develop skills that are of lasting value. Alwyn Noronha's work in Goa is a prime example.

Computers are scarce in Goa. When Noronha started on his Goa School Computers Project there were just 720 computers for schools that had 110,000 pupils. Since January 2002 Noronha's project, organised by a group of expatriate Goan software-programmers, has provided hundreds of recycled computers for local schools. The trouble was not the hardware, which they managed to get almost for free, but the software. A licence for Microsoft Windows would have cost $60 per computer. Paying a local company to support machines running Microsoft programs is expensive and Noronha did not have a maintenance budget. He wanted the children and their teachers to be able to maintain the computers themselves: that way they would learn how they worked, and maintenance costs would be zero. That kind of self-help is impossible with Windows because the source code is closed, and Noronha wanted something that could be adapted to the local languages of Goa. So Noronha installed

Linux and other open-source programs (as well as using some Microsoft products, such as Office, because the government IT syllabus demands it). Goa's three Linux user groups support the schools for free. The computer science teachers became their own programmers, swiftly followed by their most able pupils. Noronha created computer centres in scores of rural schools using recycled hardware and open-source software at virtually no cost. Multiply the impact of that many, many times, across schools in Asia and Africa, and it will provide a new model of knowledge-sharing for development.

That is why more developing nations are adopting policies to promote open-source, a wave that began in 2003 with the election of President Lula's Workers' Party in Brazil. The province of Rio Grande do Sul had approved a Bill for Free Software Development in December 2002, but after Lula's election the Ministry of Technology started to fund open-source software development and training for public officials to implement it. By 2005, seven of 22 federal ministries were using open-source, and a presidential decree was drafted making it compulsory for federal bodies to adopt open-source. Brazil has produced a wave of open-source projects including Connectiva, a Brazilian version of Linux. Both the voting system and bank ATMs run on open-source. The Brazilian army and statistics body have both adopted open-source, as have the state-owned Banco do Brasil, postal service and oil company. São Paolo's public Internet cafés, Telecentres, all run open-source. Not everything has gone smoothly: the education ministry bought 12,000 computers for schools to run the Open Office programme only to find there was no version in Portuguese. When the Workers' Party was re-elected in 2006 the commitment to open-source was watered down, tempting proprietary software companies to sell at lower prices.

Other developing nations and the world's largest computer companies are watching Brazil's experiment closely. Open-source is cheaper. In 2005, for every computer used in public administration Brazil paid Microsoft $500 in licensing fees. By then Brazil's statistical agency had not updated its software for five years because, it said, it could not afford to do so. The government estimates it will save at least $120 million a year by switching to open-source, in a country where public debt is 50 per cent of GDP and 30 per cent of the population live below the poverty line. Gilberto Gil, the lean and charismatic guitar-playing culture minister, argues that open-source makes sense to a generation of political leaders brought up with the 1960s Brazilian counter-culture of Tropicalismo, which blended music from around the world. Gil sees Tropicalismo and open-source as creative responses to the flow of ideas made possible by globalisation: 'It is a cannibalistic response of swallowing what they gave us, processing it and making it something new and different.' Sérgio Amadeu, the architect of Brazil's open-source strategy, argues that reliance on proprietary software creates a dependency culture whereas open-source allows people to build up their own skills. Jonathan Schwartz, chief executive of Sun Microsystems, in April 2006 put the impact this way in a blog-posting following a visit to the country:

> Brazil is one of the more progressive nations in the world when it comes to the use of free and open-source software. It's got one of the largest and most vibrant developer communities.

More developing countries are likely to follow in Brazil's wake. In September 2005, the Peruvian Congress passed a bill prohibiting any public institution from buying systems that tie users into software limiting their autonomy. Venezuela's

President Hugo Chavez mandated in 2004 that the government switch to open-source. In Argentina, the government's pragmatic policy promotes competition between open and closed software. A 2004 survey found that almost 50 per cent of major businesses in Argentina were using open-source.

However, as we have already noted, the real significance of open-source will be in Asia, where policy-makers view open-source as a way to challenge US dominance in computing.

On a freezing day in Seoul in early 2006 I found myself sitting in the warm office of a senior policy-maker at the South Korean Ministry of Information and Communications who was responsible for the country's much-vaunted strategy to make broadband ubiquitous. The country's rapid economic development since the end of the Korean War – when 55 per cent of the population were illiterate and only a few hundred people a year graduated from university – is the product of a deeply conservative culture which has made innovation a national cause. The one area in which the official felt South Korea had made progress in co-operating with its traditional rivals, Japan and China, was promoting open-source software:

> If we look at what we have managed to achieve in mobile phones, Samsung and other South Korean companies are on a par with the world leaders. That is because the technology for mobile phones is very open, based on open standards. That has allowed us to innovate. But we cannot do that in personal computers so much because of the control that the US has. We will have no chance of creating the next generation of personal computers unless it's based on open-source software, probably Linux. That is the only way, long-term, for us to break the dominance of Microsoft.

Over the next decade, as millions of people in India and China are lifted out of poverty they will connect to the Internet. They will be unwilling to pay high prices for US software from the likes of Microsoft. Their governments will want to comply with the rules and regulations of world trade and will not want to be seen to be condoning widespread illegal copying. The way out may well be to create open-source platforms for shared but legal low-cost development.

South Korea offers a glimpse of how We-Think could shape the future in Asia, particularly China. OhmyNews, with just 55 employees, orchestrates the work of 55,000 'citizen reporters' who submit their articles and opinions on any aspect of current affairs. Between 80 per cent and 90 per cent of South Koreans in their twenties have a 'minihompy' (mini homepage), an online space for photo-sharing, keeping a journal and networking. In Seoul it is common to see couples sitting together in Internet cafés playing massive multi-player online games. Open-source accounts for 26 per cent of the software used in government. Where South Korea is now, the prosperous regions of mainland China will be within 10 years.

Finally, We-Think's style of organisation is particularly well suited to the developing world, where professionals are in short supply, and centralised, top-down solutions will not work in far-flung villages. The social enterprises of the developing world are prime candidates for taking up these self-help technologies to address the needs of the poorest. We-Think-style solutions may not always take off in the developed world because they will have to compete with entrenched, industrial-era organisations. The developing world which has yet to establish many of these institutions might provide much more fertile ground.

The ideas to guide such collaborative, self-help solutions are already at work, for example in the Barefoot College created by Bunker Roy in 1972. Roy had turned his back on life in a wealthy Delhi family with the aim of creating an institution that would give India's illiterate villagers greater control over their lives, by helping them to provide heat, light, clean water and food for themselves. He could not afford to employ professionals to teach the villagers, and city-based experts had little understanding of village life anyway. So Roy trained a small group of villagers to become teachers and engineers. They taught others, who in turn became engineers, teachers and doctors in their own villages. By 2007 two generations of some families had become professionals, thanks to the college. In the villages around the college, each evening more than 4,000 children who tend cattle by day attend night classes with teachers in education centres lit by solar-powered lanterns installed by engineers. They drink clean water from one of the more than 1,737 hand-operated water pumps that have been installed since 1979 and are maintained by 1,200 mechanics, and which provide water for more than 325,000 people. In Roy's world, demand can generate its own supply: people who need light can become lighting engineers; learners can become teachers.

The most famous example of barefoot thinking in action is Bangladesh's Grameen Bank, founded in 1976 by Muhammad Yunus, an economics professor, to provide very poor people with micro-credit. Traditional banks, reliant on professional expertise, regarded poor people seeking loans as unprofitable. Grameen employs a small body of professionals who train an army of barefoot bankers who then work with village committees to administer Grameen's tiny loans. By 2003, Grameen had lent more than $4 billion to about 2.8 million Bangladeshis, including in 570,000 mortgages to build tin-roofs for huts to keep people dry during the monsoons. Grameen is 90 per cent owned by the people it lends to. Grameen's motto might be We-Think so We-Bank.

The barefoot, self-help model is being replicated by social entrepreneurs across the developing world. In India, Jeroo Billimoria built up a national telephone service for street children, almost entirely by training children to advise one another. The campaigning organisation Witness – which works under the banner 'See it, film it, change it' – provides human-rights groups around the world with video technology, editing support and advocacy to tell their stories directly to the mainstream media. It helps to turn victims of human-rights abuses into media producers and storytellers. In the Mbuya parish of Kampala, the capital of Uganda, Margrethe Junker, a Belgian nun, is leading an Aids support network, for more than 1,350 clients, with just 230 volunteers, 77 per cent of whom are clients of the service themselves. When I met her in 2005, she told me,

> We had no option. We had such huge need and no doctors, we just had to do it by organising people to do it themselves. The more people become involved as contributors, the better they feel.

Professionals create a distribution bottleneck, which is why the most imaginative social innovations in the developing world, such as Grameen, employ barefoot models of organisation.

What will happen when these bottom-up, self-help models are combined with low-cost communications technologies that allow mass collaboration? One glimpse of the future is M-PESA, a micro-finance pilot in Kenya, which uses mobile phones to connect borrowers and lenders. Only 1.3 per cent of Kenyans have Internet access and there are only 300,000 land-line telephones in the country, mainly in government offices. Only 10 per cent of people, mainly in towns, have bank accounts. Yet mobile-phone networks cover 70 per cent of the country and by 2007 some 6.5 million people had mobile phones, up from just 1 million in 2000. In 2004 Vodafone's Kenyan affiliate, Safaricom, and the UK government's Department for International Development (DFID) each invested about £900,000 in M-PESA (*pesa* means money in Swahili) which allows someone to use their mobile phone like a credit card or bank account. An M-PESA member can go to a mobile-phone airtime provider, usually a local shop, and upload some credit from their telephone service providers. They can use the credit themselves or transfer it to another user directly, without going through the bank. Kenya has few banks but lots of airtime dealers. Overnight M-PESA created a low-cost banking infrastructure that can be used from dawn till midnight, ideal for a highly distributed and poor population who transfer small sums: in the pilot phase of the project the average transfer between users was $4.50. In February 2007 Vodafone announced a joint venture with Citibank to take the M-PESA model worldwide, targeting transfers from the world's 190 million migrant workers back to their families, transfers worth in total about $268 billion a year.

The reach of services using mobile phones could be vast. In 1996 there were 15 million telephone lines in Africa. In 2004 Africa added 15 million new mobile-phone subscribers, and a report commissioned by DFID estimates that by 2010 there could be 200 million. In India, the government's 2010 target is 50 million subscribers. In 2005 there were 1.4 billion mobile-phone users in developing-world markets; by 2010 that could be 3 billion. Hundreds of millions of poor consumers will not be able to pay for high-cost, professional solutions to their needs for information, education, banking and health. They will be much more open to shared, collaborative solutions that blend the network and the village, the geek and the peasant.

The slow-moving, top-heavy industrial models of organisation that developed in Europe and the US in the 20th century will not work in these fast-growing but low-income economies. They instead go for low-cost solutions such as Grameen and M-PESA that mobilise participants in their millions.

Will We-Think be good for equality? Yes.

Freedom

In Thomas More's *Utopia*, which is more a warning of the risks of living in an ideal society than a blueprint for one, there are no police because the citizens keep an eye on one another. Critics of the web such as Andrew Keen, author of the polemic *Cult of the Amateur*, allege that this is exactly what the web is creating: a user-generated police state, in which everyone keeps track of everyone else. In the US a social-networking site now allows people to knit together information from published sources – addresses, the electoral role, business listings – to create maps that show who lives in which house in an area and what they do. As so often there is less new and threatening in this than people think: 19th-century maps of London listed who lived in which house and what their occupations were.

Nevertheless, people worry that every action they take – going too fast in their cars, straying into a bus lane, bunking off work, mixing ordinary waste with their recycling – could be tracked, recorded and brought back to haunt them, because every move anyone makes on the web leaves a little electronic

trail that can be tracked. With mobile-phone cameras everyone can become a paparazzo or, worse, a snooper. Photo-sharing sites such as Flickr can recognise any photo with your face in it, no matter who it was taken by, and tag it automatically. No one can be sure of anything remaining private. People need to be able to keep large parts of their life beyond the public gaze to retain a sense of dignity.[5] The Soviet Communist regimes robbed citizens of their dignity by constant eavesdropping and minor intrusions into what should have been private life; now the Internet may be achieving something similar in our culture, with our willing but sometimes naïve participation. All those embarrassing pictures and revelations teenagers are making on social-networking sites will come back to haunt them. In the autumn of 2007, two young British professional tennis players had their funding withdrawn after revealing on their Facebook sites how much they enjoyed a good night out with their mates. The web could be what Alessandro Acquisti and Ralph Gross of Carnegie Mellon University call an 'eternal memory of all our indiscretions'. Worse, it could be a busybody's charter. If the web corrodes our privacy it seems it must be bad for freedom.

A closely related fear is that younger generations are growing up with a shrunken sense of individuality, unable to think for themselves until they know what everyone else in their social network is thinking. All too quickly We-Think can become group-think as people blindly follow the herd. The web could enforce conformity rather than encouraging individuality. Jaron Lanier, in a widely read online essay published in May 2006, alleged that 'digital Maoism' was promoting collective stupidity. People were taking their lead, Lanier argued, from the all-wise 'collective' rather than bothering to think for themselves. His case is only strengthened by web advocates, such as Kevin Kelly, the original editor of *Wired*, who claim that the web is creating a 'hive mind', that of an anonymous collective in which individuals are like bees or ants. The American sociologist Sherry Turkle has echoed these fears, questioning whether young people will form a sharp sense of individual identity if they are tethered to their phone and social relationships, unable to be alone and to reflect on what matters to them because they are acting up to the online audience that constantly accompanies them.

The critics of We-Think consider it will promote more social surveillance and reduce the scope for privacy, increase group pressures on people to conform and limit the scope for individuality. Freedom cannot be extended if privacy is limited and individuality shrinks. If the critics are right, the web and We-Think will be bad for freedom.

Freedom is a slippery idea, but I believe that the web will be good for freedom of expression in four respects. These are: the freedom to think what we like, to form and express ideas independently; the freedom to shape our identities, to be who we want to be; the freedom as consumers to choose and buy what we want; and the freedom to express ourselves through creating things that matter to us.

As we have seen, while old-style industrial media consigned us to merely watching and reading, the web vastly extends the range of people who can join public debate and expands the range of the ideas they can propose. How we consume information, what news and views we get, is fundamental to how we see the world and make decisions. The cultural activities we engage in and the interests we pursue have a huge bearing on who we think we are, what matters to us, the vantage point from which we view the world. True, on the web as elsewhere in life we cluster around topics we care about and seek out people who care about similar issues. But that does not make us intellectual lemmings because on the web we have to read, listen, talk, link and debate with others, voicing our views and having them modified and changed by others. The collective products of the sometimes conversational, sometimes raucous debate that the web enables are valuable only if they draw together many independent individual contributions. They would be dull in the extreme if everyone had the same view. What draws us back are both the subjects that we care about and the passions we share with other people *and* the possibility that we might learn something new from someone with a point of view different from our own.

Google's search engine aggregates the independent judgements of millions of people. It does not ask people to submerge themselves into a collective Google identity. No one has to join a local Google cell to learn to think in the correct way, which is what Lanier's critique alleges. It is true that what people think is heavily influenced by what other people think. But the Internet did not create peer pressure, even if it may have made us more aware of others' opinions.

The few empirical tests of whether social networking is producing more conformist thinking have yielded equivocal results. In one, conducted in 2004 by Duncan Watts, a sociologist at Columbia University, and two of his students, 14,000 people were recruited through a social-networking site, divided into groups and asked to rate music played by unknown bands. As the experiment unfolded participants were given more information about other people's judgements. Often songs that started off slightly more popular became even more popular as people within the group became more aware of what others were thinking. But it was impossible to predict which groups would choose which songs. Within any one group people eventually converged on the same song; but between groups there was considerable diversity. The answer seems to be that as long as there are many groups for us to be a part of there will continue to be a huge plurality of ideas, even if within those groups there may be consensus.

So far as freedom of thought is concerned, we are no worse off with the web, and probably better off. We are more able to voice our views, to find others that share them and to learn how to modify them while being more aware of what others think. If the web meant we took our cue from the herd rather than thinking for ourselves, then Lanier would be right to worry. So long as We-Think encourages a plurality of groups to form, and those groups are formed by people making independent judgements, then it should be good for freedom of thought.

Fears that young people are losing their sense of identity because they spend so much time online with others also seem overdone. That is the conclusion reached by Danah Boyd of the University of California and other young sociologists who have studied social networks extensively. Boyd has immersed herself in social networks since the late 1990s. Her account is that teenagers regard them as largely safe, public places where they can establish who they are, gossip, flirt, bitch about their parents and sound off without fear of being interrupted by adults. Once teenagers might have done this face to face, at the bus station on the way home from school, or hanging out in a park. Yet as those real-world social spaces have come to be seen – rightly or wrongly – as riskier than they were, so teenagers do more socialising online. Many adults believe that naïve young teenagers are pursued through social-networking sites by paedophiles and identity thieves – and clearly there are risks of this kind. But young people are also learning to look after themselves and one another online. When I raised this with my 12-year-old son he pointed out that he goes into social networks only to talk to people he already knows. If a stranger came along he would soon be spotted. As Boyd put it,

> These fears are so painfully overblown. Is there porn on MySpace? Of course. And bullying, sexual teasing and harassment are rampant among teenagers. It is how you learn cultural norms and roles. These kids need to explore their lives among strangers. Teach them how to negotiate this world. They need a place that is theirs.

The young seem better equipped for a world where some of their lives are on semi-public display than their elders. More young people are archiving their adolescence in real time, putting more information about themselves into the public domain, warts and all. Older people are uncomfortable with much of this activity; they understand neither how nor why young people do it. They dismiss the young for mistaking virtual friends for real ones; for sending illiterate text messages; for being easily diverted and being happy only when connected to multiple screens and communication devices at once. Yet much of this online activity is done regardless of whether there is an audience for it. Most blogs attract zero comments: they are audience-less media. Most young people blog and post pictures online to feed their own sense of satisfaction; only a minority make a spectacle of themselves to earn notoriety. Most of the teenagers Emily Nussbaum talked to in her investigation of social networking, for an article in *New York Magazine* on 12 February 2007, seemed healthy and normal – albeit with thicker skins than their parents and a distinctly social sense of their identity. As one told her, 'What is the worst that is going to happen? Twenty years down the road someone's gonna find your picture? Just make sure it's a good picture.'

By the time the current generation of teenagers grows up, everyone will be revealing a part of themselves online. It will be the new normal. In South Korea, almost all university students have a profile on a social-network site: not to have one is considered odd. Social networks are a new way for young people to establish their sense of identity, rather than having it swamped. They are more likely to find out who they are through interaction with other people their age than through monastic reflection in their bedrooms.

As Charles Taylor, the Canadian philosopher puts it,

> My discovering my identity doesn't mean that I work it out in isolation but that I negotiate it through dialogue, partly overt, partly internalised, with others. That is why the development of an ideal of inwardly generated identity gives a crucial importance to recognition. My own identity crucially depends on my dialogical relations with others.[6]

Many of the younger, social-networking generation would sign up to Taylor's idea of identity and freedom. They are collaborative individualists – they have a strong sense of their individual rights, ambitions and aspirations but they have grown up being highly social, thanks to the web and mobile phones. When my stepdaughter Henrietta organises her friends with text messages they are like an electronic herd grazing across north London until they converge on the same bar. Worries that they are over-socialised are wide of the mark, especially as most critics of the new economy, such as Richard Sennett, complain of exactly the opposite: that young people are too individualistic and atomised. If Sennett is right, we need more social tethering, not less. The web is certainly changing how young people establish their sense of individuality; it is not extinguishing it.

Our freedom as consumers has also grown, not only because we can now search more easily across many more products but also because the digital economy allows a far greater diversity of products to co-exist. Chris Anderson, the editor of *Wired* magazine, argues in his book *The Long Tail* that in more industries a few hit products that reach millions of consumers will trail behind them a very long tail of many products that reach niche markets of a few consumers. Anderson's long-tail thesis emerged from examining the economics of Netflix, the online movie-rental business, which makes a lot of money from millions of people renting a few blockbusters and almost as much again from a few people each renting hundreds of movies. When this long tail of mini-markets is added together it is as large as the mass market. Serving the long tail can be profitable, Anderson argues, because the web now makes distribution of products to far-flung markets much cheaper. A town of 600,000 people is too small to be an economically viable market for a long tail of cult videos. In a town this size, niche markets will be too small for it to be economic for suppliers to provide for them. But if the Internet can expand that market to 6 million or even 60 million, then the chances of finding enough consumers who might want a slightly oddball movie increases. Sustaining low-margin niche markets is easier when the consumers do some of the work themselves. The We-Think combination of community, participation and niche markets lies at the heart of one of the most successful businesses of the last decade: eBay.

There is nothing new about eBay: it has taken the flea market global by using the Internet to connect sellers to an unimaginably large pool of buyers. As a start-up eBay could not afford to have a centralised customer-services department; by default the business had to be built with a bottom-heavy structure in which communities of collectors traded both goods and tips helping one another to work out how to make best use of the trading platform. That core of early traders provided the norms that now shape the much larger, more diffuse range of participants that eBay has since attracted. eBay the company is built on the scale and self-organisation of eBay the community. As eBay's understated, Canadian first president Jeff Skoll, acknowledged when I spent an afternoon with him in San Jose in 2005,

> Traditional companies did try to copy us but each time they did, they just treated people like wallets. They did not see them as a community of participants, and so they did not really connect with them. They just saw the users as transactions to be managed. People did not stay loyal to eBay because the technology was better but because they wanted the sense of community.

eBay the company created a shared trading platform, laid down some simple rules, made it very easy for participants to take part and then provided them with tools – such as the item-for-sale form – that allowed them to get on with it themselves. The rating system, through which buyers rank sellers and sellers can build up a reputation for reliable customer service, is a simple form of We-Think: a rating system based on hundreds of individual judgements that allows the market to function with minimal central control for quality. This ethic of share-and-do-it-yourself is what makes eBay so low-cost and because it is low-cost, it can sustain a mass of niche markets where people can trade what they want. On eBay you can buy everything from a Barbie doll to parts for a Rolls-Royce. Even Wal-Mart's site cannot match that range.

There are limits to how far the long tail can spread. There is no long tail in steel, oil, water, electricity, telephone services, aerospace and defence. Even where there *is* a long tail, smaller producers making cult products for tiny niche markets struggle to keep their heads above water. There may be little money to be made in the long tail unless, like Netflix, you can aggregate those small pockets of demand. Yet the combination of community and the market pioneered by eBay and Craigslist has created more ways for consumers to find a wider range of products. The Internet is good for consumer freedom. That, however, is the least important dimension of the freedom the web expands: what really counts is the way the web expands the freedom to undertake creative work.

More people enjoyed the freedom to be consumers in the 20th century. The freedom to express oneself through creative work, however, remained the preserve of a few people doing jobs that were specially designated 'creative'. For most people, work is still a necessity; people get their sense of self-realisation from their leisure. The spread of tools for creativity, participation and collaboration is changing that. Every day, millions of people use computers, software, cameras and other equipment to create pictures, music, films, games, animated cartoons and web pages, to connect to an audience and to find collaborators. On YouTube can be found the finely crafted videos of Lasse Gjertsen, a young self-taught film-maker from Larvik in Norway, which have been viewed more than 2 million times. On Flickr are the haunting photographs of a young

Icelandic photographer that were later used in advertisements for Toyota. There will be more Internet-born stars like the comics Ze Frank and Ask a Ninja. MySpace is a treasure-trove of music by unsigned bands. Software like Sibelius allows someone to play a keyboard connected to a computer and see the notes transcribed onto a score. With a few more clicks, the melody can be orchestrated for a symphony orchestra or a rock band. On the Sibelius community website, pro-am composers publish, sell and share their work. Within a few months of its creation Sibelius had 45,000 member-contributed scores and 20 new scores arriving every day. Jeremy Silver, Sibelius's chief executive – who runs perhaps the only company in the world that has to give software-programmers time off to play in orchestras – explained,

> Sibelius is really a tool for the extension of the imagination and we want to take that to as many people as possible. The fact that the professionals use it matters a lot to the product's standing, but the real impact comes when it spreads to hundreds and thousands of people, especially children. Then it could transform how they can be creative, together.

More tools for creativity are on their way. By 2017, teenagers will not be merely loading photographs onto websites and penning blogs; they may be running their own Internet television channels, with content they have created and borrowed from one another. Making an animated film could be as easy as putting together a PowerPoint presentation.

The experience of being creative, expressing ourselves through what we make rather than what we buy, will become available to more people, at least for part of their lives. Critics deride this as a dumbing-down of culture, a corrosion of professional quality. Yet in the decades to come a mass, digital folk culture will emerge, in which people will create, borrow, share, adapt and imitate one another. That will make our economy and society better. Edmund Phelps, the 2006 Nobel prize-winner in economics, argues that opportunities to do creative work should define a 'good' economy:

> Of huge importance is the economy's capability of providing people with prospects of careers generating mental stimulation, intellectual challenge, problem-solving and maybe the exercise of creativity, thus prospects of personal development (self-realisation) and various attainments (independence, recognition and pride in earning one's way).

The big opportunity to promote more self-realisation in the already materially rich developed economies is not in building more malls and extending opportunities for consumption but in opening up more opportunities to do creative, satisfying work. In the 20th century, the mass of people in the developed world gained the freedom to be consumers at the price of becoming waged workers in the iron cage of industrial organisations. In the 21st century more people will get some experience of being free to create, thanks to the web.

The significance of these opportunities to be creative was brought home to me in 2006 when I met Marysia Lewandowska, a Polish-born, London-based artist who had recently investigated amateur film-making clubs in Communist Poland, where in the 1970s and 1980s the authorities had encouraged film-making to distract people from the lure of morally corrupting US entertainment. Film-making clubs based in factories had made scores of feature-length films, many of them to high production standards. Factory hands had directed, acted, made sets, edited and played the music. Many years later Lewandowska interviewed many of those who'd been involved in the films to find out what they got from the experience. Her conclusion was: 'They were learning how to be free. To think for themselves and express themselves.'

The web should extend the freedom for people to form, express, share and debate a plurality of ideas. The older generation's worries that socially networked young people are unable to shape their own identities seem misplaced. The web is extending greater freedom of choice to consumers by making markets more diverse, especially through low-cost, eBay-like business models that sustain a long tail of niche markets. The most important change that We-Think brings, however, is to extend the freedom to be creative. That experience of creative and productive freedom, for so long confined to just a few special people working in special places, is being made much more widely available. More people will learn how to be free by expressing themselves.

Will We-Think be good for freedom? Yes.

We-Think has huge potential to be good for democracy, equality and freedom. So I conclude it should be good for us, on condition that we can make the most of it. That will happen only if we acknowledge that although we live in market economies, what we choose to share and hold in common will be as vital to our future as what we keep for ourselves, and that will be especially true for ideas.

7
AS WE MAY THINK

An idea is set in motion by being shared.[1] The range of tools available for pooling, exchanging and developing ideas determines the extent of our possible innovation and creativity and so fundamentally our prosperity, well-being and hope for the future. Ideas grow by being articulated, tested, refined, borrowed, amended, adapted and extended, activities that can rarely take place entirely in the head of a single individual; invariably they involve many people sharing different insights and criticisms. The web allows shared creativity of this kind to involve more people, discussing more questions from more angles with more ideas in play – at least it does as long as people organise themselves in the right way. This book reports on some of the first, crude efforts at large-scale, shared creativity. There will be many more such efforts, large and small, some of which will fail, before we have a better sense of what amateurs and professionals, users and producers can achieve when they work together in the right way. Whatever that might be we will not discover it overnight; it will involve a real struggle against powerful vested interest, conventional wisdom and scepticism, both warranted and not; there will be costs as well as benefits. Cherished institutions and familiar ways of working will be threatened along with the privileged role of professional, authoritative sources of knowledge.

The idea that sharing might become the economy's motive

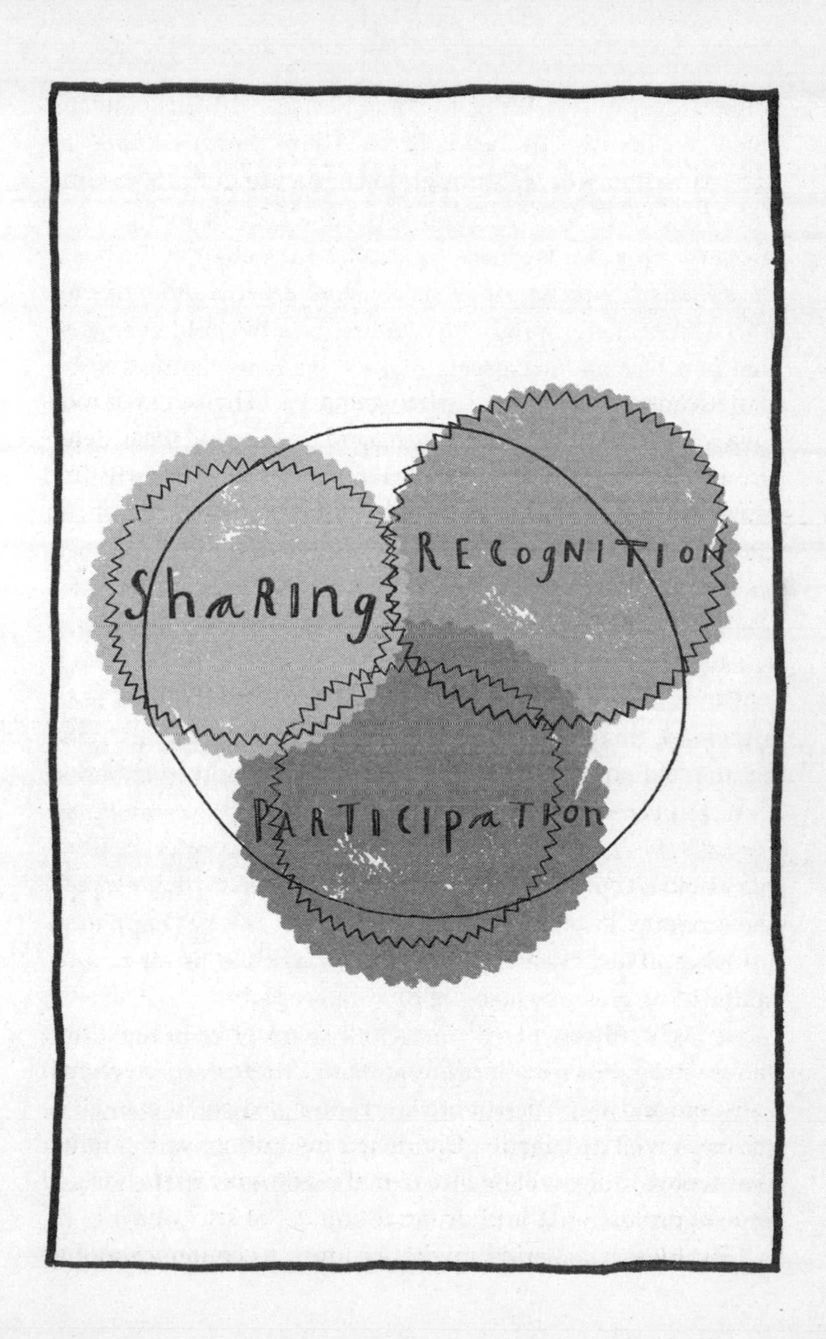
Sharing
RECogNITION
PARTICIPaTIon

force is deeply unsettling to some because it turns conventional wisdom on its head. From Adam Smith's *Wealth of Nations* written in 1749 through to Hernando de Soto's recent *The Mystery of Capital*, economists have argued that private property provides the basis for capital – the elixir at the heart of capitalism, which propels its constant growth and renewal. Karl Marx called capital 'the hen that lays the golden eggs' – that part of a nation's assets that allows new ventures to be initiated and productivity increased. A house, parcel of land, piece of machinery or factory becomes capital only if it is fixed in a tangible form, as property, so that it can be used as collateral – to be borrowed against or notionally divided so others can invest in it. The distinction between assets and capital is a tricky one, but assets are a stock of wealth – like reserves of oil – whereas capital is more like a flow of energy the oil creates when it is refined into petrol.

The charismatic Peruvian economist Hernando de Soto argues on this basis that much of the world remains poor because although many people have savings, land and houses, these assets have not been fixed as property and so cannot play the role of capital. The considerable assets of poor economies have not been put into a form – private property – that propels the economy into ceaseless motion, which is what happened in the capitalist West. The answer, for de Soto, is for poorer countries to create systems of private property.

De Soto's thesis has attracted its share of criticism from those who point out, on the one hand, that property rights cannot be created without law and good government and, on the other, that property rights are not worth that much unless the owners have the capacity to do something with their land. My own reservation is more basic still. De Soto, following in Adam Smith's footsteps, might be right in contending that private property animates the industrial and agricultural economy of land and things. But in an economy of ideas, much – and perhaps most – of the animating capital is shared.

The web's significance is that it makes sharing central to the dynamism of economies that have hitherto been built on private ownership.

That is why the new organisational models being generated by the web are so unsettling for traditional corporations created in an industrial model of private ownership. Traditionally organised and owned corporations are seeking to *exploit* a platform that animates both ideas and markets by allowing sharing on an unprecedented scale. Life will be uncomfortable for many mainstream organisations and the people who work in them because the web, the most potent modern force for innovation, encourages a culture of *sharing* that is deeply at odds with the idea that private property animates our economies. Many of the early experiments in We-Think seem so odd because the participants give away what they create. They treat ideas and knowledge, at least some of the time, as if they were gifts.

The web reconnects us with a different story about the rise of the West: one that gives a central role to the way ideas are aired and shared rather than focusing on how land and buildings are locked down into property. Much of what we most value – in culture, language, art, science and learning – comes from a kind of gift exchange, in which ideas are passed from person to person, and accumulate over long periods. The cultural experiences we most value – a Maxim Vengerov performance, a glimpse of a Terracotta Warrior, a Barbara Hepworth sculpture – cannot be measured by the entry price we pay. The impact and value of ideas and culture – their capacity to touch and move us, to make us think and imagine

– live on well after the experience we paid for is over. We may pay to go to a museum but when we leave we carry something with us that might last for ever. That is the gift component of the experience. Almost every exchange of ideas has this dual character: even when money changes hands there is a gift and, once given, a gift passes beyond the control of the giver; it cannot be contained. Gifts are neither transactions nor acts of charity; they create bonds and forge relationships between people, convey emotions and invite reciprocation. Financial transactions thrive in open markets; gifts thrive in relationships and, when the gift is an idea, by circulating through a community. As Lewis Hyde puts it in his history *The Gift*, 'When gifts circulate within a group their commerce leaves a series of interconnected relationships in its wake and a kind of decentralised cohesiveness emerges.'[2]

The more we value activities that depend on the gift of ideas – science, culture, art, learning and innovation – the more we will need non-market institutions to support them, as well as markets to exploit them. In the past, the Church, the monarchy, the aristocracy and, more recently, the state have supported these non-market activities of shared inquiry, expression and experimentation. We-Think creates a new basis for the gift economy, in which thanks to decentralised and distributed technologies gifts of knowledge and ideas can circulate from and to many people. That circulation will sustain the mass, digital folk culture that will be the most powerful cultural force of the first decades of the 21st century. Global public banks of knowledge, Wikipedia, the Encyclopedia of Life, the Human Genome, the outputs of the International Polar Year, the free content on The Sims all belong in part to this commons.

This is not to suggest that everything can be done through non-market means. Far from it: the market brings freedom, choice and higher productivity. We need the right balance between market and non-market ways of organising the networked economy. The civil war that erupted in the Homebrew Computer Club between Bill Gates, who wanted to own ideas, and Fred Moore, who wanted them to grow by being freely shared, will be resolved only if neither side completely wins. The most exciting organisational models of the future will mix collaboration and commerce, community and corporation. These efforts may well start with gifts – like Linus Torvald's donation of his core programme to create the Linux community – which eventually provide a platform for a mass of commercial activity – for companies like Red Hat which install Linux for corporations. They can also start the other way around, with a transaction – someone paying for the software that allows them to start playing World of Warcraft – and develop to enable a mass of gifting – players sharing content with one another in self-organising guilds.

The industrialised economies spent most of the 20th century honing the management disciplines of mass production, from industrial relations to branding and supply-chain management. At the start of the century there was no science of management. By its end we had business schools, expensive executive training programmes, libraries of books, and pricey management consultancies. The web's creative potential will be realised in the century to come only if we become equally sophisticated about managing and leading creative collaborations. In theory the technology should help: new generations of search tools will make it easier to scour for information using trails others have already blazed; simulation technologies currently available only to architects and engineers will allow us to visualise and communicate problems and possible solutions; new technologies for collaboration will make it much easier for people to work creatively across borders and disciplines on shared projects and in shared virtual spaces.[3]

So the road to prosperity in a world shaped by the web will lie in the interaction between gifts and transactions, sharing and owning; between markets that trade products and communities that breed knowledge. Successful organisations will have a foot in both camps. The most successful will find ways to make money from enabling the circulation of ideas. As innovation becomes more central to the way we make our livings and how we tackle pressing challenges we face – from global warming to health pandemics – our well-being will depend more and more on what we share with others and create together.

That said, it is also clear that We-Think will not be sustained by Utopian ethics. If We-Think has to depend on altruism it will not last long. Collaborative business models succeed because they reward people, satisfy desires, achieve personal objectives: they get a job done for people who want to share photographs, play games, find information, write software, sign up to a cause, unlock a gene, follow a faith, borrow some money, link to friends. Yet these practical benefits are underpinned by something even more powerful and less instrumental: a desire for recognition.

People are drawn to the experiments in We-Think because they seek recognition for the value of the contribution they can make to a shared endeavour. They want their worth to be recognised by people who count – their peers. Will Wright, the designer of The Sims, believes that a desire for recognition attracts people to mass computer games. The same is true of participants in open-source software projects, scientific communities and Wikipedia – just as it was for 18th-century

Cornish engineers who disclosed their engine designs to one another.

Our desire to be acknowledged and accepted, paid attention and respected is deep-seated, as Adam Smith argued in his other great work *The Theory of Moral Sentiments*, an account of the ethics that should accompany a liberal, market society:

> What are the advantages which we propose to gain by the great purpose of human life which we call bettering our condition? To be observed, to be attended to, to be taken notice of with sympathy, complacency and approbation, are all the advantages we can propose to derive from it.

As societies get richer and more of the basic needs for food, clothing, housing, warmth and security are met, people will become increasingly interested in the psychological dimensions of well-being. It is vital to our psychological well-being that we are held in esteem, valued and recognised for what we do. Our identities – what we are good at and what matters to us – depend on the recognition of other people. In the past, certainly in the rich world, many people acquired a sense of identity from their position in a bounded local community. In the 20th century, occupation and position in an organisational hierarchy often provided the key. Now, people increasingly get a sense of identity from the relationships they form and the interests they share with others. The web matters not least because by allowing people to participate and share, it also gives them a route to recognition, if only through the comments posted in response to a blog, a rating as a trader on eBay, the points acquired as a game player, or the incorporation of software they have written into the source code. People are drawn to share, not only to air their ideas, but in the hope their contributions will be recognised by a community of their peers.[4]

Recognition cannot bought and sold on the market – at least not the kind of recognition that counts. The recognition of sycophants and fawning underlings is not worth much. Many people, perhaps mainly men, get a sense of status from the position they occupy in a hierarchy. But those positions have become far less secure. Celebrity, job titles and even wealth can come and go; personal regard and recognition need deeper roots to be valuable. Those roots cannot be found from within: self-regarding people are usually slightly unpleasant. We win the recognition that counts from impartial, external sources, usually communities of our peers. That is why We-Think culture is so powerful: the communities and social networks that the web is spawning are a vital way for people to obtain recognition from their peers for what they do, especially if that involves ideas. These communities meet a basic human need that will get stronger as we become materially richer.[5]

Ideas are animated when they are shared, and people are driven to share because recognition and regard can be reliably earned only from communities, networks, clans, families, religious groups, movements that are not animated by money. The social web provides people with a new way to win recognition for being a good player, programmer, film-maker, singer, composer, citizen, writer, scientist, researcher and so on. Our well-being depends on our being esteemed by people we ourselves hold in high esteem. So long as the web continues to provide a way for people to earn recognition it will continue to grow.

However, this sets up a further potential conflict, however, relating to the terms on which people will be allowed to participate in these activities and how. Groups are wise, clever and smart only when they are made up of independent people who are capable of thinking for themselves and armed with diverse skills and points of view. Sydney Brenner pulled together such a group for his worm project, as did the players in I Love Bees. We will be able to create durable, reliable complex goods like encyclopaedias, software programs, guides to designing and making products, maps and scientific theories only if we find ways of working together that multiply our shared intelligence. Some collaborations will involve only professionals: scientists, for example, will find it easier to collaborate with large numbers of other scientists on more complex subjects, over greater distances. Other collaborations will be more like Wikipedia: assemblies of amateurs, some rather ramshackle, others – like OhmyNews in South Korea – very well organised. Many of the most successful – like modern astronomy – will combine professional and amateur knowledge rather than setting them at odds.

Feeding this development will be a fundamental change in our economic culture: over the next two or three decades people will start to play quite different roles, seeing themselves increasingly as participants and contributors, as well as workers and consumers. In the 20th century almost everyone in the industrialised world was a worker and a consumer in the mass-production economy: a worker by day for a wage, a consumer at night and at the weekends. By the end of the 21st century, our grandchildren are likely to see themselves more as participants, contributors and innovators, if only in modest ways. More people than ever will have access to tools that will allow them to create words, images, films, pictures, music, software, objects, machines, perhaps even organisms. Armed with technologies that embody and mimic professional intelligence, more amateurs will be able to do work of higher quality. As technology gets cheaper and software more intuitive, powerful tools will flow out of scientists' laboratories, doctors' surgeries and designers' studios into bedrooms, living rooms and classrooms, enabling dedicated, knowledgeable and ambitious amateurs to do more work more creatively. Activities that once had to be centralised because they were costly and complex, such as manufacturing and even energy supply, may become more distributed and decentralised. Consumers in the rich developed world may not take up Gershenfeld's Fab Labs or Bowyer's RepRap, but in the poorer developing world they well might.

Participation will, however, mean quite different things in different settings. More companies and brands, politicians and celebrities will try to incorporate their consumers as fans and followers, recruiting celebrants. They will participate, but more in the way a congregation does in a church service. Fundamentalists and terrorists are using the web in a different way, to connect widely distributed followers, disciples and adherents of a faith: the people involved in these networks are not just followers but also activists and initiators. Hacker communities, different again, bring together self-governing, democratic communities of digital craft workers. The participation enabled by the web will have a wide range of applications.

It would be naïve to imagine that a new way of organising ourselves will necessarily be exclusively positive. There will be downsides, possibly very significant ones – while industrial mass production massively increased productivity and brought cheaper goods within the reach of most people, it has also been accompanied by alienation and strife at work, industrial accidents, ravaged landscapes and environmental despoliation on a vast scale. So will We-Think culture have similarly damaging downsides? What will we lose that we value, and what new dangers will present themselves?

As we have seen, critics are already warning us to worry about a whole slew of possible disadvantages: the erosion of professional authority and knowledge; the loss of individuality in a morass of social networking; the eradication of spaces for reflection as a result of our being constantly connected; and the degradation of friendship when relationships are mediated by technology.

Strengths often breed matching weaknesses: the web's power comes from the way it allows people to share, and that may also be its greatest flaw. There are no central gatekeepers to control access to the Internet. It is a platform that virtually anyone can join, on which they can find other people, connect with them and start to share ideas. A network originally designed to allow an élite of technically savvy US academics and researchers to share files has blossomed to embrace hundreds of millions of people, each with their own reason to want to join in myriad activities. And, unlike television, the Internet has allowed people to adapt the technology as they use it, so generating yet more applications. The Internet's remarkable culture of openness – no one in particular owns it and virtually anyone can join – will be the source of our greatest challenges and risks. For sharing can also spread diseases, infections and viruses. Ideas and identities once stolen can be spirited away and spread across thousands of computers.

On the evening of 2 November 1988, Robert Tappan Morris, a Cornell University graduate student, sent a small piece of software from his university to a computer at MIT. When the software started running on the MIT machine it started to seek out other nearby computers connected to the Internet and behaved as if it were a person wanting to log on.[6] The software's sole purpose was to propagate by cracking its way into each new computer. By the following morning between 1,000 and 6,000 computers had been affected by what was the first computer virus. As Jonathan Zittrain, professor of Internet governance and regulation at Oxford University, puts it, nearly everything one needs to know about the risks to Internet security reside within this story. By 2003, a researcher connected to the Internet a computer that seemed to have been left open for others to connect to. It took spammers just 10 hours to find the machine and to start sending mail through it. Within three days, the computer had recorded 229,468 distinct messages directed at 3,360,181 would-be recipients. Open networks make us vulnerable. In May 2007, the Internet infrastructure of Estonia, one of the most connected societies in Europe, ground to a halt as its public and commercial organisations were hit by an attack from more than 1 million computers co-ordinated from Russia. Chinese hackers are busy working their way into computer systems across the West; employees at some Western defence companies have had Chinese viruses personally targeted at them that would give the hackers access to vast quantities of precious data.

So the biggest challenge we will face by far will be how to retain a semblance of control when powerful technologies are seeping out of the hands of responsible institutions and professionals into society at large, possibly to groups where there is little respect for intellectual property or good governance. In their different ways, all the web's critics converge on a single worry: it makes the world more unreliable, threatening and out of control. Whatever the limitations of top-down, industrial-era institutions, at least the world they created was relatively orderly and people knew where they stood.

Editors, academics, doctors, scientists and professionals were the gatekeepers of knowledge; they could be trusted to tell us what was fact and what not. Instead of generating more knowledge, the web often seems to sow more doubt and uncertainty, spreading speculation and gossip. The Cold War might have perched the world on the edge of nuclear annihilation, but at least the technologies of mass destruction were in the hands of powerful states that could be held accountable for their use. Now the means to spread terror are available to anyone with a camcorder, and deadly biological weapons could probably be made using ingredients bought on eBay with information gleaned from Wikipedia. You do not have to believe in conspiracy theories to worry. By 2050, tools for genetic engineering could be available to hapless amateurs releasing dangerous mutations from greenhouses equipped with gene sequencers.

In the world of the web, sharing is the goose that lays the golden eggs. Our challenge is to create a sense of order and security without undermining our capacity for sharing. There will be three main ways to bring this potentially unruly world under control.

First, those who have top-down control will fight to retain it, even as power threatens to seep away from them. The greatest struggle will be in China, where there will be a drawn-out wrestling match between the entrenched, hierarchical, authoritarian power of the Communist Party and the swelling power of an increasingly autonomous civil society. Only slightly less dramatic and probably equally prolonged will be the attempts of software, entertainment and media companies to control how what they call 'content' is used and paid for, a battle they will fight through rights-management technologies and legal moves. Many of them recognise that they are swimming against a current that is only getting stronger. The chairman of one of the largest publishing groups in the US told me over dinner in London in late 2007,

> We have decided that the price we will be able to charge for our content is in effect the cost of copying it. Soon that is all people will be prepared to pay. We will have to make more of our money from other services, not just publishing.

Within organisations, managers and professionals will struggle to retain power based on privileged access to information and knowledge as the people they lead and serve become less deferential. Followers will increasingly acquire their own voices, challenge official sources and look for their own information. One of the most pressing questions will be when we will want top-down control by professionals – for example to control flows of dangerous technologies – and when not.

Secondly, there will be more forms of peer-to-peer control, and even surveillance, that work with the grain of a world in which information and power are more decentralised and distributed. In industrial-era media the quality of information was controlled by professional editors and regulators. As publishing is distributed to many hundreds of millions of people, top-down control of quality will not work. Instead there will have to be more open, transparent, peer review and rating. We will get used to rating one another and being rated by our peers. Scientists are kept honest by open, peer review. We will need something similar in many other walks of life.

Thirdly, we will have to encourage more self-control so people use their growing technological power responsibly. That means, at the very least, children learning the skills and norms of media literacy and responsibility; learning to question and challenge information as well as copy and paste it. We will have to learn how to feel secure, while sharing more of our ideas and identity with other people. This will mean that where there is top-down control it will have to be more intelligent, transparent and focused, and if we want more to be done through these self-governing networks we will have to accept being more accountable to our peers.

Today more people, organisations, markets and cities are more connected than they have ever been. What passes between them – goods, information, ideas, money, diseases – travels at higher speed, like cars speeding down a four lane highway, so that the potential for a catastrophic pile-up is huge. Most of our worries about the world that is opening up to us come back to the fact we have little option but to share with people we do not know and cannot necessarily trust. Our growing connections to other people also leave us exposed to them. We are unavoidably implicated in and compromised by far-off events: in the summer of 2007 panicked British savers queued in their thousands to withdraw their money after a downturn in part of the distant US mortgage market; the threat of al-Qaeda terrorists incubated in the border zones of Afghanistan and Pakistan casts a shadow over people's lives in London and Madrid; Aids is killing more people than any previous pandemic. Keeping things stable when so many people can be connected to so much so easily takes ever more effort, whether in the form of financial market regulation or of the millions of interconnected decisions we will need to make to tackle climate change. Studies of forests show that as they grow they become more diverse, with plants and animals increasingly occupying specialised niches, to make the most of all the available nutrients. Yet after a while forests can become so densely connected that they become increasingly vulnerable to an external shock, such as fire, which can spread very fast. The world being created by the web is rather like a fast-growing forest – increasingly dense, diverse and connected. When the web really is ubiquitous, will it become more like a forest that is so densely interconnected that a small fire can quickly sweep through vast areas of woodland? Or will it continue to find its equivalent of new sources of water, soil and sunlight to feed continued growth?

Much will depend on whether We-Think culture can rise to the challenges that are facing us. Mass production came of age during the fight against Fascism in the Second World War. We-Think might come of age in the fight against global warming, because finding alternative ways to generate energy, use resources and cope with rising sea-levels will require collective innovation on an immense scale. In the Netherlands, the sea is kept at bay only thanks to an intricate system of dikes, dams, pumps and sluices, requiring everyone to rely on one another or else drown. The Netherlands exists only through cumulative, communal innovation which has built a country out of reclaimed swamp and sea. The Dutch tend not to laud individual, superstar designers. Instead they focus on evolutionary, practical innovations, such as simple modular buildings that can be easily adapted. The Netherlands is We-Think in action at national level: a constant, adaptive, interconnected and incremental approach to innovation.[7] In future we will need this kind of highly social innovation at much greater scale if we are to tackle global issues.

As the Canadian catastrophe theorist Thomas Homer-Dixon puts it in *The Upside of Down*,

> In one respect humanity is extraordinarily lucky: just when it faces some of the biggest challenges in its history, it has

> developed a technology that could be the foundation for extremely rapid problem-solving on a planetary scale, for radically new forms of democratic decision-making.

We have only just begun to tap the web's potential and the new ways of thinking and acting it offers us. We-Think will really make a difference when we use it creatively to tackle major shared challenges: to spread democracy and learning, to improve health and quality of life, to tackle climate change and the threats of extremism. If we succeed in bending it to those objectives, people might look back a century from now and say it made the critical difference in the world's ability to govern itself. We-Think tells a new story about how the global knowledge economy could develop, offering a way to create new generations of shared public goods for software, education, communications, health and food production. If globalisation is to be no more than the march of McDonald's, Coke and Microsoft, it will be a shallow and distorted account of what Western culture has to offer that many in the developing world will reject. We-Think offers a different possible story, one of trust and collaboration built on liberal and enlightenment traditions of peer collaboration in pursuit of better ideas, arbitrated on the basis of evidence rather than ideology. We-Think will spread these values through civil society far more effectively than machines dispensing Coke. We-Think is a tool not just for economic development but also for democratic and social development.

We are compelled to share our ideas; that is how they come to life. And when we share ideas they multiply and grow, forming a powerfully reinforcing circle. You are not defined simply by what you own. You are also what you share. That should be our credo for the century to come.

EPILOGUE: THINK *WITH*

The culture the web is creating can be reduced to a single, simple design principle: call it the principle of *With*.

The web invites us to think and act with people, rather than for them. The web is an invitation to connect with other people with whom we can share, exchange and create new knowledge and ideas. The principle that we should think 'with' stands in stark contrast to the kind of outlook, organisation and culture spawned by the mass production and mass consumption of the twentieth century.

The truth is that often in the name of doing things for people, traditional and hierarchical organisations end up doing things to people. Companies say they work for consumers but in reality treat them like targets to be aimed at, with wallets to be emptied. The person who calls himself my 'personal relationship manager' at a leading high street bank does not know me from Adam, but in the attempt to sell me some savings products I do not want pretends that we are lifelong friends, on first name terms. Most people's experience of public services is no better. Social services departments are designed to help people in need, but those on the receiving end often complain that they feel they are treated as a number rather than an individual, that their care and support is closely rationed by an impersonal, complex process that rarely attends to what people really need. Similarly, politicians claim they are working for us, on our behalf, representing our views. But, most of the time they seem to be spinning messages and broadcasting at us. In schools, all too often, instruction is delivered to you, as you listen to your teacher or copy from the blackboard. Even in hospitals it feels as if you are being processed by the system. We live with systems that are meant to treat us like consumers but which are impersonal, rigid, inhuman even, whether in the public or the private sector.

And of course often people feel the market – that great abstract force in people's lives – is acting upon them, uprooting jobs, industry or community rather than working for them. Work in many large organisations feels like an imposition. Too much of management feels like it is making people do things they do not really want to do.

These common and widespread experiences of being manipulated stem from deeply rooted assumptions. Knowledge and learning flows from specially designated experts to people in need. Organisations are hierarchies based on the power and the knowledge to make decisions. Centralised authority is exercised top-down. The industrial system extracts resources from one place, changes them into a commodity or product and delivers them in changed form to waiting consumers. Knowledge is largely instrumental and rational: it allows us to master, plan and control our environment.

The web, however, is creating a world that works to the logic of *With* – structured lateral, free association of people and ideas. The principle of doing things with people rather than to or for them will breed very different organisations, services and experiences in virtually every field.

With is about how we work. I dropped out of working for large organisations because I wanted to be able to work alongside people, without hierarchies and job titles getting in the way. The working culture of open source communities, such as Wikipedia, and the web more generally, is a culture in which ideas are shared with like-minded people. In the world of the web you can freely communicate with anyone you need to regardless of title or hierarchy. And this principle is at the heart of great social enterprises such as the Grameen Bank and the Barefoot College, who identify people's problems and devise solutions with them, thereby building capabilities that allow people to go on and sustain themselves. They have the same peer-to-peer, do-it-yourself spirit as the new organisations being created on the web.

Driven by creative collaboration and shared conversation the principle of *With* is central to innovation. *With* should be the guiding principle of politics in liberal communities, where politicians should be working with people to find a solutions to shared problems. People want a more conversational form of politics, where their options are sought and heard rather than being spun messages or broadcast to from on high. The spirit of *With* played out recently in the American elections, with thousands of volunteers campaigning on the web, encouraging voters to help Obama on his way to the White House. The concept of *With* means looking at our relationship with our physical environment in a different way and creating an economy that works inconjunction with the environment. Efforts to recycle resources, to minimise waste production and pollution released into our environment are all examples of *With* working in practice.

Relationships are vital to our well being. The difference between a life that feels rich and full, and one that feels empty and hollow, often lies in the quality of our relationships: whether we feel significantly connected to others. Relationships and networks are the basic building blocks of society. Learning too is enhanced when it is collaborative and interactive. An active participant, someone who learns with other people rather than passively from them, is more likely to develop their knowledge in the long run.

The twentieth century was dominated by big organisations that did things for us and to us as workers and as consumers. Could the twenty-first century be about organisations that work with us and allow us to do things by ourselves? I think it could be, but for that possibility to be realised we will have to meet three big challenges:

First, who really participates, or to put it another way, who are you with? Is this collaborative culture just for the ultra-connected, activists, fans and hobbyists, or a wider population?

Second, how do you increase participation to make it more meaningful? (Participation can be achieved in many ways, whether by leaving a comment on a webpage or making a proposal, uploading a video, writing some sharable software, or donating money. All of these actions give feedback to other users.)

Third, can we do what the Levellers failed to do – create robust and reliable ways for people to collaborate and share?

If the collaborative web turns into just another inspiring but doomed experiment then it will have been a failure. To make a lasting change the logic of *With* needs to infiltrate everyday life, people need to be able to rely upon it to get what they need, as workers and consumers.

Over the next few years we will bear witness to a struggle between two forces; the familiar, but dysfunctional world in which decisions are made for us and where actions are done in our name on our behalf versus the emerging, illusive and potentially revolutionary world in which we think and work with one another.

The idea of *With* has transformative potential. If you want one very simple way of thinking your way into the world the web is creating, think *With*.

ACKNOWLEDGEMENTS

In writing *We-Think* I have drawn on the ideas of many other people, with or without their knowledge. Several books heavily influenced my thinking, including Ilkka Tuomi's *Networks of Innovation*, Steven Weber's *The Success of Open Source*, Yochai Benkler's *The Wealth of Networks*, Henry Chesbrough's *Open Innovation*, Steve Johnson's *Emergence*, Carliss Baldwin and Kim Clark's *Design Rules*, Eric von Hippel's *Democratizing Innovation*, C. K. Prahalad and Venkat Ramaswamy's *The Future of Competition*, Scott Page's *The Difference*, James Surowiecki's *The Wisdom of Crowds* and Chris Anderson's *The Long Tail*. The other books, papers and articles I have drawn on are listed in the bibliography. Should you think that is incomplete, you can go to the wiki version of the book and add in your own references and links.

Many of the ideas were developed through projects I worked on with other people or by speaking at seminars and workshops. Those people include: Carol Coletta at Ceos for Cities in Chicago; Cathy Brickman and her team at Virtuel Platform, Joeri van Steenhoven at Knowledge Land, Frans Nauta and Michiel Schwartz, all in Amsterdam; Cecilie With and everyone else involved with InnoTown in Ålesund in Norway; John Hartley, Stuart Cunningham and their team at Queensland University of Technology in Brisbane; Jimmy Wales and others at Wikimedia Foundation; Rowena Young and Anthony Hopwood of the Saïd Business School; Valerie Hannon and the team at the Department for Education's

Innovation Unit; Ed Miliband, Dominic Maxwell, Ben Jupp and Campbell Robb at the Cabinet Office and the Office of the Third Sector; David Miliband and Ravi Gurumurthy at the Foreign and Commonwealth Office; James Wilsdon, Paul Miller, Molly Webb and Kirsten Bound, my colleagues in the Atlas of Ideas project at Demos; Tom Bentley, formerly of Demos and now a government adviser in Melbourne; John Craig, chief executive of the Innovation Exchange; Geoff Mulgan at the Young Foundation and Tom Steinberg at My Society; Hilary Cottam, Colin Burns and Jennie Winhall, my colleagues at Participle; Jonathan Kestenbaum and Richard Halkett at the National Endowment for Science, Technology and the Arts where I am a visiting fellow; Kirstin Undheim, trend analyst at Opinion Bengal in Oslo; Polly LaBarre and Bill Taylor, authors of *Mavericks at Work*; Tom Kenyon Slaney and everyone else at the London Speaker Bureau for helping to arrange most of my speaking engagements; and Ed Mayo, Philip Cullum and Sue Johnston at the National Consumer Council for sponsoring my work on user-driven innovation. Thanks to James Cherkoff of Collaborate Marketing for alerting me to the joys of Arseblog.

The book could not have been finished without the efforts of my researcher Anna Maybank, then freshly graduated, who produced detailed, comprehensive and insightful reports at incredible speed on everything from World of Warcraft to Brazil's open-source policy. Anywhere you see an interesting fact it will have come from Anna. Thanks to Mike Bond and Martin Coyne for introducing me to Andy Headington and Alex Othold at Adido for designing the book's website and Debbie Powell, who did the book's brilliant illustrations in the weeks after she had graduated from Bournemouth College of Art. Freddie Norton, my stepson, and Josh Booth, then an intern at Demos, did sterling work to pull together the bibliography, notes and the A–Z of We-Think.

I would like to thank all the people who contributed online by leaving comments on my site and by emailing me; including Heiko Spellek, Miranda Mowbray, Tim Sullivan and the others too many to mention, some of whom are highlighted in the preface.

My agent, Clare Alexander, immediately saw the project's potential and handled it with her inimitable touch. Andrew Franklin and Daniel Crewe at Profile also saw the project's potential and showed considerable patience when it took me much longer than I had promised to get them a decent draft. I am glad to be published by Profile because they care so much about the books they publish.

My greatest thanks are reserved for my family. I owe the title to a discussion over the breakfast table on holiday with my sister-in-law, Elaine Bedell, living proof that brains and looks do go together. I had to write much of the book early in the morning at weekends when my family often wanted me to be doing something more entertaining and helpful. Sometimes they could not disguise their shock that I was still writing *that* book, the one I had been writing months before, and I could see their point. So thanks to Henrietta, Freddie, Harry and Ned. My greatest debt, however, is to my wife, Geraldine Bedell, who never wavered in her support and in the final stages of writing added a new dimension to my writing: grammar. I would not have been capable of writing it without her.

NOTES

Prologue

1 Funtwo's video can be found at http://uk.youtube.com/watch?v=QjA5faZF1A8. Accessed on 5 November, 2008 it had received 51,221,981 hits. Many other classical electric guitar videos have had several million hits.

2 For more on the likes of charlieissocoollike see Celia Hannon, Peter Bradwell, Charlie Tims, Video Republic, Demos, 2008. http://www.youtube.com/user/charlieissocoollike

3 *The World Turned Upside Down*, Christopher Hill, Penguin, 1972

Chapter 1

1 Thomas Homer-Dixon (2006), *The Upside of Down: Catastrophe, Creativity, and the Renewal of Civilisation* (Souvenir Press Ltd, 2007)

2 Jane McGonigal, 'Why I Love Bees: A Case Study in Collective Intelligence Gaming', February 2007. Available from http://www.avantgame.com/McGonigal_WhyILoveBees_Feb2007.pdf

3 In April 2007, for example, about 4,000 flash mobbers took over the main concourse of Victoria Station in London, armed with personal stereos, to dance for two hours, in the middle of the afternoon. See Howard Rheingold, *Smart Mobs* (Perseus Books, 2002)

4 Sanger's work was funded by a company called Bomis, of which Wales was one of the directors.
5 Larry Sanger, 'The Early History of Nupedia and Wikipedia', in Chris DiBona, Danese Cooper and Mark Stone (Eds), *Open Sources 2.0* (O'Reilly, 2006)
6 The most famous example was when an article alleged that a former aide to Robert Kennedy was involved in President John F. Kennedy's assassination.
7 For a detailed account of this case see Yochai Benkler, *The Wealth of Networks* (New Haven, CT/London: Yale University Press, 2006)
8 David Edgerton, 'From Innovation to Use: Ten (Eclectic) Theses on the History of Technology', *History and Technology* 16 (1999), pp. 1–26. Originally published in French as 'De l'innovation aux usages. Dix thèses éclectiques sur l'histoire des techniques', *Annales HSS* 4–5 (1998), pp. 815–37; David Edgerton, *The Shock of the Old: Technology in Global History since 1900* (Profile Books Ltd, 2007)
9 F. B. Viégas, M. Wattenberg and K. Dave, 'Studying Cooperation and Conflict between Authors with History Flow Visualizations', CHI (2004), pp. 575–82. Summarised in William Emigh and Susan C. Herring, 'Collaborative Authoring on the Web: A Genre Analysis of Online Encyclopaedias', 2005. Available from http://ella.slis.indiana.edu/~herring/wiki.pdf
10 Charles Taylor, *Sources of the Self: The Making of the Modern Identity* (Cambridge University Press, 1992); René Descartes, *Discourse on Method and the Meditations*. Translated by F. E. Sutcliffe (Penguin Books, 1976)

11 Pierre Levy and Robert Bonomo (trans.), *Collective Intelligence: Mankind's Emerging World in Cyberspace* (Perseus Books, 1997)

Chapter 2

1 Tim O'Reilly, 'What is Web 2.0?', *www.oreillynet.com*, November 2005. Available from http://www.oreillynet.com/pub/a/oreilly/tim/news/2005/09/30/what-is-web-20.html
2 Matthew Gray, 'Web Growth Summary', *www.mit.edu*. Available from http://www.mit.edu/people/mkgray/net/web-growth-summary.html
3 Mark Brady, 'Blogging: Personal Participation in Public Knowledge-Building on the Web', Chimera Working Paper, University of Essex, 2005. Available from http://www.essex.ac.uk/chimera/content/pubs/wps/CWP-2005–02-Blogging-in-the-Knowledge-Society-MB.pdf
4 Rebecca Blood, 'Weblogs: A History and Perspective', *Rebecca's Pocket*, September 2000. Available from http://www.rebeccablood.net/essays/weblog_history.html
5 Mallory Jensen, 'Emerging Alternatives: a Brief History of Weblogs', 2003. Available from http://www.cjr.org/issues/2003/5/blog-jensen.asp
6 http://portal.eatonweb.com
7 http://www.technorati.com/about
8 http://slashdot.org and http://www.digg.com
http://www.plastic.com
http://www.fark.com
9 See Anna Maybank, 'Web 2.0', at *www.charlesleadbeater.net*.
10 See http://english.ohmynews.com

11 Nicole Ellison, Charles Steinfield and Cliff Lampe, 'Spatially Bound Online Social Networks and Social Capital: The Role of Facebook', Department of Telecommunication Information Studies and Media, Michigan State University, 2006. Available from http://msu.edu/%7enellison/facebook_ica_2006.pdf
12 Danah Boyd, 'None of This Is Real: Identity and Participation in Friendster', University of California, Berkeley. Available from http://www.danah.org/papers/NoneOfThisIsReal.pdf
13 http://c2.com/cgi/wiki?WikiHistory
14 The Economist New Media Survey, 'The Wiki Principle', *The Economist*, April 2006. Available from http://www.economist.com/surveys/displaystory.cfm?story_id=6794228
15 See Steven Levy and Brad Stone, 'The New Wisdom of the Web', *Newsweek*, April 2006. Available from http://www.msnbc.msn.com/id/12015774/site/newsweek
16 Fred Turner, *From Counterculture to Cyberculture* (Chicago, IL/London: University of Chicago Press, 2006)
17 Patrice Flichy, *The Internet Imaginaire* (Cambridge, MA: MIT Press, 2007)
18 Charles Leadbeater, 'The DIY State', *Prospect* 130, January 2007
19 Fred Turner, op. cit.
20 John Markoff, *What the Dormouse Said: How the Sixties Counterculture Shaped the Personal Computer Industry* (Penguin, 2006)
21 Patrice Flichy, *The Internet Imaginaire* (Cambridge, MA: MIT Press, 2007)
22 Jonathan Lethem, 'The Ecstasy of Influence', *Harper's Magazine*, February 2007

23 Garrett Hardin, 'The Tragedy of the Commons', *Science* 162 (1968), pp. 1243–48
24 Elenor Ostrom, *Governing the Commons* (Cambridge University Press, 1990)
25 Lawrence Lessig, *Code and Other Laws of Cyberspace* (New York, NY: Basic Books, 1999) and *Free Culture* (New York, NY: Penguin Press, 2004)
26 Melvyn Bragg, *The Routes of English* (BBC Factual and Learning, 2000); Melvyn Bragg, *The Adventure of English* (Hodder & Stoughton Ltd, 2003)
27 Jonathan Lethem, 'The Ecstasy of Influence', *Harper's Magazine*, February 2007
28 Cory Doctorow et al., 'On "Digital Maoism: The Hazards of the New Online Collectivism" By Jaron Lanier', *Edge* (2006). http://www.edge.org/discourse/digital_maoism.html
29 Paul A. David, 'From Keeping "Nature's Secrets" to the Institutionalization of "Open Science"', in Rishab Aiyer Ghosh (Ed.), *Code* (Cambridge, MA/London: MIT Press, 2005)
30 Alessandro Nuvolari, 'Open Source Software Development: Some Historical Perspectives', Eindhoven Centre for Innovation Studies Working Paper 03.01 (2003); Koen Frenken and Alessandro Nuvolari, 'The Early Development of the Steam Engine: An Evolutionary Interpretation Using Complexity Theory', Eindhoven Centre for Innovation Studies Working Paper 03.15 (2003)

Chapter 3

1 Andrew Brown, *In the Beginning Was the Worm* (Pocket Books, 2003)

2 Eric S. Raymond, *The Cathedral and the Bazaar* (O'Reilly, 2001)
3 Doc Searls, 'Making a New World', in Chris DiBona, Danese Cooper and Mark Stone (Eds), *Open Sources 2.0* (O'Reilly, 2006)
4 Glyn Moody, *Rebel Code: Linux and the Open Source Revolution* (Penguin, 2002)
5 Like many radical innovations Linux is not as revolutionary as it first seems. Computer scientists and engineers had been sharing equipment and code for decades. Richard Stallman, a computer scientist and hacker, had started work on an open-source operating system in the mid-1980s and created the General Public License (GPL) in 1985, which allowed users to copy a program, modify it and sell versions, so long as they made their modifications freely available to others. Linux is a version of Unix, a program created in the 1960s and built on Minix, a program designed to be a teaching aid. Linux started to take off in January 1992 when Torvalds released the second version under Stallman's innovative GPL. Within days the mailing list for Linux activists had 196 members.
6 Ilkka Tuomi, *Networks of Innovation: Change and Meaning in the Age of the Internet* (Oxford University Press, 2002)
7 See http://counter.li.org/
8 David A. Wheeler, 'More than a Gigabuck: Estimating GNU/Linux's Size', *www.dwheeler.com*, July 2002. Available from http://www.dwheeler.com/sloc/redhat71-v1/redhat71sloc.html
9 Juan José Amor-Iglesias, Jesús M. González-Barahona, Gregorio Robles-Martínez and Israel Herráiz-Tabernero, 'Measuring *Libre* Software Using Debian 3.1 (Sarge) as

a Case Study: Preliminary Results', *UPGRADE* 6.3, June 2005. Available from http://www.upgrade-cepis.org/issues/2005/3/up6-3Amor.pdf

10 Steven Weber, *The Success of Open Source* (Cambridge, MA/London: Harvard University Press, 2004)

11 Thomas Kuhn, *The Structure of Scientific Revolutions* (University of Chicago Press, 1962), p. 10

12 Richard K. Lester and Michael Piore, *Innovation: The Mission Dimension* (Cambridge, MA/London: Harvard University Press, 2004)

13 Andrew Hargadon, *How Breakthroughs Happen* (Boston, MA: HBS Press, 2003)

14 James Surowiecki, *The Wisdom of Crowds* (Little, Brown, 2004)

15 Scott E. Page, *The Difference: How the Power of Diversity Creates Better Groups, Firms, Schools, and Societies* (Princeton University Press, 2007)

16 Bart Nooteboom, *Learning and Innovation in Organizations and Economies* (Oxford University Press, 2000)

17 Steven Weber, *The Success of Open Source* (Cambridge, MA/London: Harvard University Press, 2004)

18 Articles by Lakhani, Ghosh and Lerner in Joseph Feller, Brian Fitzgerald, Scott A. Hissam, Karim R. Lakhani (Eds), *Perspectives on Free and Open Source Software* (Cambridge, MA: MIT Press, 2005)

19 Josh Lerner and Jean Tirole, 'The Simple Economics of Open Source', NBER Working Paper W7600 (2000). Available from http://www.nber.org/papers/w7600

20 Robert Wright, *Nonzero* (Abacus, 2001)

21 Carliss Y. Baldwin and Kim B. Clark, *Design Rules* (Cambridge, MA/London: MIT Press, 2000)

Chapter 4

1 Richard Sennett, *The Culture of the New Capitalism* (New Haven, CT/London: Yale University Press, 2006)

2 Mitch Kapor, *blog.kapor.com*

3 Henry Chesbrough, Wim Vanhaverbeke and Joel West (Eds), *Open Innovation: Researching a New Paradigm* (Oxford University Press, 2006)

4 John Hartley, 'Culture Business and the Value Chain of Meaning', *The New Economy, Creativity and Consumption – A Symposium* (Brisbane: Queensland University of Technology Publications, 2002), pp. 39–46

5 http://www.blizzard.com/inblizz/profile.shtml

6 Nicolas Ducheneaut, Nicholas Yee, Eric Nickell and Robert J. Moore, 'Alone Together? Exploring the Social Dynamics of Massively Multiplayer Online Games', *Conference on Human Factors in Computing Systems*, 2006, p. 3. Available from http://www.parc.xerox.com/research/publications/files/5599.pdf

7 David Barboza, 'Ogre to Slay? Outsource it to Chinese', *New York Times*, December 2005. Available from http://www.nytimes.com/2005/12/09et/chnology/09gaming.html?ex=1291784400&en=48a72408592dffe6&ei=5088

8 Dominic Rushe, 'Fantasy Game Turns Internet Into Goldmine', *The Sunday Times*, September 2006

9 See http://www.ige.com

10 See http://www.worldofwarcraft.com

11 Christian Luthje, Cornelius Herstatt and Eric von Hippel, 'The Dominant Role of "Local" Information in User Innovation: The Case of Mountain Biking', MIT Sloan School of Management Working Paper No. 4377–02, July 2002. Available from http://userinnovation.mit.edu/papers/6.pdf; Eric von Hippel and Georg von

Krogh, 'Open Source Software and the Private-Collective Innovation Model: Issues for Organization Science', *Organization Science* 14.2 (2003), pp. 209–23; Eric von Hippel, 'Horizontal Innovation Networks – By and For Users', MIT Sloan School of Management Working Paper No. 4366–02, June 2002

12 Sonali Shah, 'Open Beyond Software', in Chris DiBona, Danese Cooper and Mark Stone (Eds), *Open Sources 2.0* (O'Reilly, 2006)

13 Henry Jenkins, *Convergence Culture* (New York University Press, 2006)

14 Henry Jenkins, *Fans, Bloggers, and Gamers* (New York University Press, 2006)

15 Pekka Himanen, *The Hacker Ethic and the Spirit of the Information Age* (London: Secker & Warburg, 2001)

16 John Roberts, *The Modern Firm* (Oxford University Press, 2004)

17 Jane Jacobs, *The Death and Life of Great American Cities* (Vintage, 1992)

18 John Micklethwait and Adrian Wooldridge, *The Company* (London: Weidenfeld & Nicolson, 2003)

19 Henry Hansmann, *The Ownership of Enterprise* (Cambridge, MA: Belknap Harvard, 1996)

20 James Boyle, 'The Second Enclosure Movement and the Construction of the Public Domain', *Law and Contemporary Problems* 66.1&2 (2003), pp.33–74. Available from http://www.law.duke.edu/journals/66LCPBoyle

21 Lawrence Lessig, *Free Culture* (New York: Penguin Press, 2004)

Chapter 5

1 William C. Taylor and Polly LaBarre, *Mavericks at Work: Why the Most Original Minds in Business Win* (London: HarperCollins, 2006)
2 Sonali Shah, 'Open Beyond Software', in Chris DiBona, Danese Cooper and Mark Stone (Eds), *Open Sources 2.0* (O'Reilly, 2006)
3 Don Tapscott and Antony D. Williams, *Wikinomics: How Mass Collaboration Changes Everything* (Penguin, 2007)
4 Freeman Dyson, 'Our Biotech Future', *The New York Review of Books* 54.12, 12 July 2007
5 See http://www.bookcrossing.com
6 Timothy Ferris, *Seeing in the Dark* (New York: Simon & Schuster, 2002)
7 Chris DiBona, Sam Ockman and Mark Stone (Eds), *Open Sources: Voices from the Open Source Revolution* (O'Reilly, 1999); Philip Ball, 'Life, But Not as We Know It', *Prospect*, August 2007

Chapter 6

1 Dick Morris, *Vote.com* (Renaissance Books, 1999)
2 Joe Trippi, *The Revolution Will Not Be Televised* (New York: HarperCollins, 2004)
3 Pippa Norris, *Democratic Phoenix* (Cambridge University Press, 2002)
4 Amartya Sen, *The Unwanted Indian: Writings on Indian History, Culture and Identity* (Penguin, 2006)
5 Jeffrey Rosen, *The Unwanted Gaze: The Destruction of Privacy in America* (New York: Vintage Books, 2001)
6 Charles Taylor, *The Ethics of Authenticity* (Cambridge, MA: Harvard University Press, 1991)

Chapter 7

1 Vannevar Bush, 'As We May Think', *Atlantic Monthly*, July 1945. Available from http://www.theatlantic.com/doc/194507/bush
2 Lewis Hyde (1983), *The Gift: How the Creative Spirit Transforms the World* (Edinburgh: Canongate, 2007)
3 Mark Dodgson, David Gann and Ammon Salter, *Think, Play, Do: Technology, Innovation and Organization* (Oxford University Press, 2005)
4 Charles Taylor, *The Ethics of Authenticity* (Cambridge, MA: Harvard University Press, 1991)
5 Avner Offer, *The Challenge of Affluence* (Oxford University Press, 2006)
6 Jonathan L. Zittrain, 'The Generative Internet', *Harvard Law Review* 119. 1974 (2006)
7 Aaron Betsky, *False Flat: Why Dutch Design Is So Good* (Phaidon, 2004)

BACKGROUND RESEARCH

More detailed references can be found in the background research reports compiled by the book's researcher Anna Maybank. These detailed reports are available for download from my site www.charlesleadbeater.net. They cover the following topics:

Wikipedia: Examines the accuracy and economics of the online encyclopaedia.

OhmyNews: A briefing on the online news site.

Linux: A history of the popular open-source operating system, including an overview of the Linux community and a rundown of its expansion.

Digital Divides: The web's impact on equality.

International Polar Year: A multinational feat of collaboration in climate science.

2008 Presidential Campaign: The influence of web 2.0 and social networking on the US presidential election 2008.

Open-Source Manufacturing: Will the web hand production over to the people?

Folksonomy Stats: The rise of 'tagging' on the web.

Gary Kasparov: How this chess giant took on the collective intelligence of thousands.

Encyclopedia of Life: The project that will give every species its own webpage.

Web Growth: Charting the growth of the web over the last decade.

Mall of the Sims: Virtual shopping for virtual people.
Language as Commons: The rise and spread of English.
Human Genome Project: The collaborative venture to uncover our genetic selves.
Brazil and Open Source: Examining the state-led open-source movement in Brazil and beyond.
m-pesa: How the benefits of micro-banking were taken to Africa by money (*pesa* in Swahili) being made more mobile.
Ubuntu: The Linux-based, open-source, community-developed operating system.
Apache: The success of open-source web service.
World of Warcraft: How the virtual world created by this computer strategy game has changed the reality of over 8 million players' lives.
Web 2.0: Exploring the new face of the web, from blogs to wikis through folksonomies, RSS and social networking.

BIBLIOGRAPHY

Amor-Iglesias, Juan José, Jesús M. González-Barahona, Gregorio Robles-Martínez and Israel Herráiz-Tabernero, 'Measuring *libre* software using Debian 3.1 (Sarge) as a case study: preliminary results', *UPGRADE* 6.3, June 2005. Available from http://www.upgrade-cepis.org/issues/2005/3/up6–3Amor.pdf

Audretsch, David B., *Innovation and Industry Evolution* (Cambridge, MA/London: MIT Press, 1995)

Bak, Per, *How Nature Works* (New York: Copernicus, 1996)

Baldwin, Carliss Y., and Kim B. Clark, *Design Rules* (Cambridge, MA/London: MIT Press, 2000)

Ball, Philip, 'Life, But Not as We Know It', *Prospect*, August 2007

Barboza, David, 'Ogre to Slay? Outsource It to Chinese', *The New York Times*, December 2005. Available from http://www.nytimes.com/2005/12/09et/chnology/09gaming.html?ex=1291784400&en=48a72408592dffe6&ei=5088

Battarbee, Katja, *Co-Experience: Understanding User Experiences in Social Interaction* (Helsinki: University of Art and Design, 2004)

Battram, Arthur, *Navigating Complexity* (The Industrial Society, 1998)

Benkler, Yochai, *The Wealth of Networks* (New Haven, CT/London: Yale University Press, 2006)

Bentley, Tom, *Learning Beyond the Classroom* (Routledge, 1998)

Bessen, James, *Open Source Software: Free Provision of Complex Public Goods*, July 2005. Available from http://papers.ssrn.com/s013/papers.cfm?abstract_id=588763#PaperDownload.

Betsky, Aaron, *False Flat: Why Dutch Design Is So Good* (Phaidon, 2004)

Bhidé, Amar V., *The Origin and Evolution of New Businesses* (Oxford University Press, 2000)

Blitz, Roger, 'Costs Made Council Stick With Microsoft', *Financial Times*, 17 August 2004

Blood, Rebecca, 'Weblogs: A history and Perspective', *Rebecca's Pocket*, September 2000. Available from http://www.rebeccablood.net/essays/weblog_history.html

Boden, Margaret A., *The Creative Mind: Myths and Mechanisms* (Routledge, 2004)

Boyd, Danah, 'None of This Is Real: Identity and Participation in Friendster', University of California, Berkeley. Available from http://www.danah.org/papers/NoneOfThisIsReal.pdf

Boyle, James, 'The Second Enclosure Movement and the Construction of the Public Domain', *Law and Contemporary Problems* 66.1&2 (2003), pp. 33–74. Available from http://www.law.duke.edu/journals/66LCPBoyle

Brady, Mark, 'Blogging: Personal Participation in Public Knowledge-Building on the web', Chimera Working Paper, University of Essex 2005. Available from http://www.essex.ac.uk/chimera/content/pubs/wps/CWP-2005–02-Blogging-in-the-Knowledge-Society-MB.pdf

Bragg, Melvyn, *The Routes of English* (BBC Factual and Learning, 2000)

Bragg, Melvyn, *The Adventure of English* (Hodder & Stoughton Ltd, 2003)

Brown, Andrew, *In the Beginning Was the Worm* (Pocket Books, 2003)

Brown, John Seely, and Paul Duguid, *The Social Life of Information* (Boston, MA: HBS Press, 2000)

Bush, Vannevar, 'As We May Think', *Atlantic Monthly*, July 1945. Available from http://www.theatlantic.com/doc/194507/bush

Byrne, David, *Complexity Theory and the Social Sciences* (Routledge, 1998)

Castells, Manuel, *The Rise of the Network Society* (Malden, MA: Blackwell, 1996)

Castells, Manuel, and Pekka Himanen, *The Information Society and the Welfare State* (Oxford University Press, 2002)

Chesbrough, Henry, *Open Innovation* (Boston, MA: HBS Press, 2003)

Chesbrough, Henry, Wim Vanhaverbeke and Joel West (Eds), *Open Innovation: Researching a New Paradigm* (Oxford University Press, 2006)

Christensen, Clayton M., *The Innovator's Dilemma* (Boston, MA: HBS Press, 1997)

Clippinger III, John H. (Ed.), *The Biology of Business: Decoding the Natural Laws of Enterprise* (San Francisco: Jossey-Bass, 1999)

Cooke, Philip, and Kevin Morgan, *The Associational Economy* (Oxford University Press, 1998)

Coyne, Richard, *Technoromanticism* (Cambridge, MA/London: MIT Press, 1999)

Csikszentmihalyi, Mihaly, *Creativity: Flow and the Psychology of Discovery and Invention* (New York: HarperCollins, 1996)

Csikszentmihalyi, Mihaly, *Flow: The Classic Work on How to Achieve Happiness* (Rider, 2002)

Cukier, Kenneth, 'A Market for Ideas', *The Economist*, 22 October 2005

Dave K., Viégas F. B. and Wattenberg M., 'Studying Cooperation and Conflict Between Authors with History Flow Visualizations', CHI (2004), pp. 575–82. Summarised in William Emigh and Susan C. Herring, 'Collaborative Authoring on the Web: A Genre Analysis of Online Encyclopaedias' (2005). Available from http://ella.slis.indiana.edu/~herring/wiki.pdf

David, Paul A., 'From Keeping "Nature's Secrets" to the Institutionalization of "Open Science"', in Rishab Aiyer Ghosh (Ed.), *Code* (Cambridge, MA/London: MIT Press, 2005)

De la Mothe, John, and Gilles Paquet (Eds), *Evolutionary Economics and the New International Political Economy* (Pinter, 1996)

Dennis, Carina, 'Biologists Launch "Open Source Movement"', in *Nature* 431 (2004), p. 494

Descartes, René, *Discourse on Method and the Meditations*. Translated by F. E. Sutcliffe (Penguin Books, 1976)

De Vries, Marc J., *80 Years of Thinking at the Phillips Natuurkundig Laboratorium 1914–1994* (Amsterdam: Pallas, 2005)

DiBona, Chris, Danese Cooper and Mark Stone (Eds), *Open Sources 2.0* (O'Reilly, 2006)

DiBona, Chris, Sam Ockman and Mark Stone (Eds), *Open Sources: Voices from the Open Source Revolution* (O'Reilly, 1999)

Di Maggio, Paul (Ed.), *The Twenty-first-Century Firm* (Princeton University Press, 2001)

Doctorow, Cory, et al. 'On "Digital Maoism: The Hazards of the New Online Collectivism" By Jaron Lanier', *Edge*, (2006). See http://www.edge.org/discourse/digital_maoism.html

Dodgson, Mark, David Gann and Ammon Salter, *Think, Play, Do: Technology, Innovation and Organization* (Oxford University Press, 2005)

Dodson, Sean, 'Show and Tell Online', *Guardian*, 3 February 2006

Dravis, Paul, *Open Source Software: Perspectives for Development* (Washington, DC: InfoDev, 2003)

Ducheneaut, Nicolas, Nicholas Yee, Eric Nickell and Robert J. Moore, 'Alone together? Exploring the Social Dynamics of Massively Multiplayer Online Games', *Conference on Human Factors in Computing Systems*, 2006, p. 3. Available from http://www.parc.xerox.com/research/publications/files/5599.pdf

Dyson, Freeman, 'Our Biotech Future', *New York Review of Books* 54.12, 12 July 2007

Economist, The, 'The Wiki Principle', April 2006. Available from http://www.economist.com/surveys/displaystory.cfm?story_id=6794228

Edgerton, David, 'From Innovation to Use: Ten (Eclectic) Theses on the History of Technology', *History and Technology* 16 (1999), pp. 1–26. Originally published in French as 'De l'innovation aux usages. Dix thèses éclectiques sur l'histoire des techniques', *Annales HSS* 4–5 (1998), pp. 815–37

Edgerton, David, *The Shock of the Old: Technology in Global History since 1900* (Profile Books Ltd, 2007)

Ellison, Nicole, Charles Steinfield and Cliff Lampe, 'Spatially Bound Online Social Networks and Social Capital: The

Role of Facebook', Department of Telecommunication Information Studies and Media, Michigan State University, 2006. Available from http://msu.edu/%7enellison/facebook_ica_2006.pdf

Feller, Joseph, Brian Fitzgerald, Scott A. Hissam and Karim R. Lakhani (Eds), *Perspectives on Free and Open Source Software* (Cambridge, MA: MIT Press, 2005)

Ferris, Timothy, *Seeing in the Dark* (New York: Simon & Schuster, 2002)

Flichy, Patrice, *The Internet Imaginaire* (Cambridge MA: MIT Press, 2007)

Florida, Richard, *The Rise of the Creative Class* (New York: Basic Books, 2002)

Florida, Richard, *The Flight of the Creative Class* (New York: HarperBusiness, 2005)

Frayn, Michael, *Copenhagen* (Methuen, 2003)

Frenken, Koen, and Alessandro Nuvolari, 'The Early Development of the Steam Engine: An Evolutionary Interpretation Using Complexity Theory', Eindhoven Centre for Innovation Studies Working Paper 03.15 (2003)

Garud, Raghu, Arun Kumaraswamy and Richard N. Langlois (Eds), *Managing in the Modular Age* (Malden, MA: Blackwell, 2003)

Gawer, Annabelle, and Micheal A. Cusumano, *Platform Leadership: How Intel, Microsoft and Cisco Drive Industry Innovation* (Boston, MA: HBS Press, 2002)

Ghosh, Rishab Aiyer (Ed.), *Code* (Cambridge, MA/London: MIT Press, 2005)

Gillmor, Dan, *We the Media* (Farnham: O'Reilly, 2004)

Gladwell, Malcolm, *The Tipping Point* (Little, Brown, 2000)

Granstrand, Ove (Ed.), *Economics of Technology* (Amsterdam/London: North-Holland, 1994)

Gratton, Lynda, *The Democratic Enterprise* (Harlow: Pearson, 2004)

Gray, Matthew, 'Web Growth Summary', *www.mit.edu*. Available from http://www.mit.edu/people/mkgray/net/web-growth-summary.html

Häikiö, Martti, *Nokia: The Inside Story* (Pearson, 2002)

Hall, Peter A. and David Soskice (Eds), *Varieties of Capitalism* (Oxford University Press, 2001)

Hansmann, Henry, *The Ownership of Enterprise* (Cambridge, MA: Belknap Harvard, 1996)

Hardin, Garrett, 'The Tragedy of the Commons', *Science* 162 (1968), 1243–48

Hargadon, Andrew, *How Breakthroughs Happen* (Boston, MA: HBS Press, 2003)

Hartley, John, 'Culture Business and the Value Chain of Meaning', *The New Economy, Creativity and Consumption – A Symposium* (Brisbane: Queensland University of Technology Publications, 2002), pp. 39–46

Hayward, Peter J., *A Natural History of the Seashore* (HarperCollins, 2004)

Himanen, Pekka, *The Hacker Ethic and the Spirit of the Information Age* (Secker & Warburg, 2001)

Homer-Dixon, Thomas, *The Upside of Down: Catastrophe, Creativity, and the Renewal of Civilisation*, (Souvenir Press Ltd, 2007)

Hyde, Lewis (1979), *The Gift: How the Creative Spirit Transforms the World* (Edinburgh: Canongate, 2006)

InterAcademy Council, *Inventing a Better Future* (Amsterdam: IAC, 2004)

Illich, Ivan, *Tools for Conviviality* (New York: Harper & Row, 1973)

Illich, Ivan, *Energy and Equity* (Calder & Boyars, 1974)

Illich, Ivan, *Limits to Medicine* (Marion Boyars, 2002)
Illich, Ivan, *Deschooling Society* (Marion Boyars, 2004)
Isaacs, William, *Dialogue and the Art of Thinking Together* (Currency, 1999)
Israel, Paul, *Edison: A Life of Invention* (John Wiley, 1998)
Jacobs, Jane, *The Death and Life of Great American Cities* (Vintage, 1992)
Jenkins, Henry, *Convergence Culture* (New York University Press, 2006)
Jenkins, Henry, *Fans, Bloggers, and Gamers* (New York University Press, 2006)
Jensen, Mallory, 'Emerging Alternatives: A Brief History of Weblogs', 2003. Available from http://www.cjr.org/issues/2003/5/blog-jensen.asp
Joyce, Patrick, *The Rule of Freedom* (Verso, 2003)
Kapor, Mitch, *blog.kapor.com*
Kapor, Mitch, 'Does the Open Source Model Apply Beyond Software?', http://blogs.osafoundation.org/mitch/000815.html, January 2005
Kelley, Tom, *The Art of Innovation* (Doubleday, 2001)
Kerstetter, Jim, 'The Linux Uprising', *BusinessWeek*, 3 March 2003
Kochan, Thomas A. and Paul Osterman, *The Mutual Gains Enterprise* (Boston, MA: HBS Press, 1994)
Kotro, Tanja, *Hobbyist Knowing in Product Development* (University of Art and Design, Helsinki, 2005)
Kuhn, Thomas. *The Structure of Scientific Revolutions*, (University of Chicago Press, 1962), p. 10
Lakhani, Karim, and Eric von Hippel, 'How Open Source Software Works: "Free" User-to-User Assistance', MIT Sloan School of Management Working Paper No. 4117 (2000)

Lanchester, John, 'Should We Fear Google?', *Guardian*, 26 January 2006

Lane, Christel, and Reinhard Bachmann (Eds), *Trust Within and Between Organizations* (Oxford University Press, 1998)

Larsson, Ulf (Ed), *Cultures of Creativity* (Canton, MA: Science History Publications, 2002)

Leadbeater, Charles, 'The DIY State', *Prospect* 130, January 2007

Lencek, Lena, and Gideon Bosker, *The Beach: The History of Paradise on Earth* (New York: Penguin, 1999)

Lerner, Josh, and Jean Tirole, 'The Simple Economics of Open Source', NBER Working Paper W7600 (2000). Available from http://www.nber.org/papers/w7600

Lessig, Lawrence, *Code and Other Laws of Cyberspace* (New York: Basic Books, 1999)

Lessig, Lawrence, *Free Culture* (New York: Penguin Press, 2004)

Lester, Richard K., and Michael Piore, *Innovation: The Mission Dimension* (Harvard University Press, 2004)

Lethem, Jonathan, 'The Ecstasy of Influence', *Harper's Magazine*, February 2007

Levine, Rick, Christopher Locke, Doc Searls and David Weinberger, *The Cluetrain Manifesto* (Perseus Books, 2000)

Levy, Pierre and Robert Bonomo (trans), *Collective Intelligence: Mankind's Emerging World in Cyberspace* (Perseus Books 1997)

Levy, Steven, and Brad Stone, 'The New Wisdom of the Web', *Newsweek*, April 2006. Available from http://www.msnbc.msn.com/id/12015774/site/newsweek

Lewin, Roger, *Complexity: Life at the Edge of Chaos* (Phoenix, 1993)

Lewin, Roger, and Birute Regine, *The Soul at Work* (Orion, 1999)

Lewin, Roger, and Birute Regine. *Weaving Complexity & Business: Engaging the Soul at Work* (New York/London: Texere, 2001)

Leydesdorff, Loet, and Peter Van Den Besselaar (Eds), *Evolutionary Economics and Chaos Theory: New Directions in Technology Studies* (New York: St Martin's Press, 1994)

Loudon, Alexander, *Webs of Innovation: The Networked Economy Demands New Ways to Innovate* (Harlow: FT.com, 2001)

Luthje, Christian, Cornelius Herstatt and Eric von Hippel, 'The Dominant Role of "Local" Information in User Innovation: The Case of Mountain Biking', MIT Sloan School of Management Working Paper No. 4377–02, July 2002. Available from http://userinnovation.mit.edu/papers/6.pdf

Malone, Thomas W., *The Future of Work* (Boston, MA: HBS Press, 2004)

Markoff, John, *What the Dormouse Said: How the Sixties Counterculture Shaped the Personal Computer Industry* (Penguin, 2006)

McGonigal, Jane, 'Why I Love Bees: A Case Study in Collective Intelligence Gaming', February 2007. Available from http://www.avantgame.com/McGonigal_WhyILoveBees_Feb2007.pdf

McKelvey, Maureen, *Evolutionary Innovations* (Oxford University Press, 2000)

Mercer Management Consulting, *Audiences with Attitude* (Marsh & McLennan Companies)

Micklethwait, John, and Adrian Wooldridge, *The Company* (Weidenfeld & Nicolson, 2003)

Miller, Paul, and Paul Skidmore, *The Future of Organizations* (Demos, 2004)
Moody, Glyn, *Rebel Code: Linux and the Open Source Revolution* (Penguin, 2002)
Moore, Mark H., *Creating Public Value* (Cambridge, MA/ London: Harvard University Press, 1995)
Morgan, Gareth, *Images of Organization* (Sage, 1997)
Morris, Dick, *Vote.com* (Renaissance Books, 1999)
Myerson, Jeremy, *IDEO: Masters of Innovation* (Lawrence King, 2001)
Nalebuff, Barry J., and Adam M. Brandenburger, *Co-opetition* (HarperCollinsBusiness, 1996)
Nature magazine editorial, 'Open-source Biology', *Nature* 431 (2004), p. 491
Naughton, John, *A Brief History of the Future: The Origins of the Internet* (Weidenfeld & Nicolson, 1999)
Nelson, Richard R., and Sydney G. Winter, *An Evolutionary Theory of Economic Change* (Cambridge, MA/London: Harvard University Press, 1996)
Nohria, Nitin, and Robert G. Eccles (Eds), *Networks and Organizations* (Boston, MA: HBS Press, 1992)
Nooteboom, Bart, *Learning and Innovation in Organizations and Economies* (Oxford University Press, 2000)
Norris, Pippa, *Democratic Phoenix* (Cambridge University Press, 2002)
Nuvolari, Alessandro, 'Open Source Software Development: Some Historical Perspectives', Eindhoven Centre for Innovation Studies Working Paper 03.01 (2003)
Offer, Avner, *The Challenge of Affluence* (Oxford University Press, 2006)
Oram, Andy (Ed.), *Peer-to-Peer: Harnessing the Benefits of a Disruptive Technology* (Cambridge, MA: O'Reilly, 2001)

O'Reilly, Tim, 'What is Web 2.0?', *www.oreillynet.com,* November 2005. Available from http://www.oreillynet.com/pub/a/oreilly/tim/news/2005/09/30/what-is-web-20.html
Ormerod, Paul, *Why Most Things Fail* (Faber & Faber, 2005)
Ostrom, Elinor, *Governing the Commons* (Cambridge University Press, 1990)
Page, Scott E., *The Difference: How the Power of Diversity Creates Better Groups, Firms, Schools, and Societies* (Princeton University Press, 2007)
Petrosky, Henry, *Success Through Failure* (Princeton University Press, 2006)
Prahalad, C. K., and Venkat Ramaswamy, *The Future of Competition* (Boston, MA: HBS Press, 2004)
Prencipe, Andrea, Andrew Davies and Michael Hobday (Eds), *The Business of Systems Integration* (Oxford University Press, 2003)
Putnam, Robert D., *Bowling Alone* (New York: Simon & Schuster, 2000)
Raymond, Eric S., *The Cathedral and the Bazaar* (Cambridge, MA: O'Reilly, 2001)
Rheingold, Howard, *Smart Mobs* (Perseus Books, 2002)
Rivlin, Gary, 'Leader of the Free World', *Wired,* November 2003, pp. 153–54
Roberts, John, *The Modern Firm* (Oxford University Press, 2004)
Rosen, Jay, 'Journalism Is Itself a Religion', Production Values, Demos, http://journalism.nyu.edu/pubzone/weblogs/pressthink/2003/10/16/radical_ten.html (2006)
Rosen, Jay, 'What's Radical About the Weblog Form in Journalism?', *Pressthink,* http://journalism.nyu.edu/

pubzone/weblogs/pressthink/2003/10/16/radical_ten.html (2003)

Rosen, Jay, 'Bloggers vs. Journalists Is Over', *Pressthink*, http://journalism.nyu.edu/pubzone/weblogs/pressthink/2005/01/21/berk_essy.html (2005)

Rosen, Jeffrey, *The Unwanted Gaze: The Destruction of Privacy in America* (New York: Vintage Books, 2001)

Rushe, Dominic, 'Fantasy Game Turns Internet into Goldmine', *The Sunday Times*, September 2006

Schienstock, Gerd, and Timo Hämäläinen, *Transformation of the Finnish Innovation System: A Network Approach* (Helsinki: Sitra Reports Series 7, 2001)

Schwartz, Barry, *The Paradox of Choice: Why More Is Less* (New York: HarperCollins, 2004)

Sen, Amartya, *The Unwanted Indian: Writings on Indian History, Culture and Identity* (Penguin, 2006)

Sennett, Richard, *The Culture of the New Capitalism* (New Haven/London: Yale University Press, 2006)

Shaw, Patricia, *Changing Conversations in Organizations* (Routledge, 2002)

Spufford, Frances, *Backroom Boys* (Faber & Faber, 2003)

Stacey, Ralph D., *Complex Responsive Processes in Organizations* (Routledge, 2001)

Stacey, Ralph D., Douglas Griffin and Patricia Shaw, *Complexity and Management* (Routledge, 2002)

Stebbins, Robert A., *Amateurs, Professionals, and Serious Leisure* (Montreal/Kingston: McGill-Queen's University Press, 1992)

Steil, Ben, David G. Victor, and Richard R. Nelson (Eds), *Technological Innovation & Economic Performance* (Princeton University Press, 2002)

Sunstein, Cass, *Republic.com* (Princeton University Press, 2001)
Surowiecki, James, *The Wisdom of Crowds* (Little, Brown, 2004)
Sutton, Robert I., *Weird Ideas that Work* (Penguin, 2001)
Tapscott, Don, and Antony D. Williams, *Wikinomics: How Mass Collaboration Changes Everything* (Penguin, 2007)
Taylor, Charles, *The Ethics of Authenticity* (Cambridge, MA: Harvard University Press, 1991)
Taylor, Charles, *Sources of the Self: The Making of the Modern Identity* (Cambridge University Press, 1992)
Taylor, William C., and Polly LaBarre, *Mavericks at Work: Why the Most Original Minds in Business Win* (HarperCollins, 2006)
Teece, David J., Gary Pisano and Amy Shuen, 'Dynamic Capabilities and Strategic Management' *Strategic Management Journal*, 18.7, pp. 509–33, 1997
Thackara, John, *In the Bubble* (Cambridge, MA / London: MIT Press, 2005)
Trippi, Joe, *The Revolution Will Not Be Televised* (New York: HarperCollins, 2004)
Tuomi, Ilkka, *Networks of Innovation* (Oxford University Press, 2002)
Turner, Fred, *From Counterculture to Cyberculture* (University of Chicago Press, 2006)
Vaidhyanathan, Siva, *The Anarchist in the Library: How the Clash Between Freedom and Control Is Hacking the Real World and Crashing the System* (New York: Basic Books, 2004)
Volberda, Henk W., *Building the Flexible Firm* (New York: Oxford University Press, 1998)

Von Hippel, Eric, 'Horizontal Innovation Networks – By and For Users', MIT Sloan School of Management Working Paper No. 4366–02 (2002)

Von Hippel, Eric, *Democratizing Innovation* (Cambridge, MA/ London: MIT Press, 2005)

Von Hippel, Eric, Christian Luthje and Cornelius Herstatt, 'The Dominant Role of "Local" Information in User Innovation: The Case of Mountain Biking', MIT Sloan School of Management Working Paper (2002)

Von Hippel, Eric, and Georg von Krogh, 'Open Source Software and the Private-Collective Innovation Model: Issues for Organization Science', *Organization Science* (2003)

Von Hippel, Eric, *The Sources of Innovation* (Oxford University Press, 1995)

Waldman, Simon, 'Who Knows?', *Guardian*, 26 October 2004

Walton, John K., *The British Seaside* (Manchester University Press, 2000)

Walton, John K., *Fish & Chips and the British Working Class 1870–1940* (Leicester University Press, 2000)

Weber, Steven, *The Success of Open Source* (Cambridge, MA/ London: Harvard University Press, 2004)

Weick, Karl E., *Making Sense of the Organization* (Malden, MA/Oxford: Blackwell, 2001)

Weinberger, David, *Small Pieces Loosely Joined* (Cambridge, MA: Perseus Books, 2002)

Wheatley, Margaret J., *Leadership and the New Science* (San Francisco, CA: Berret-Koehler, 1999)

Wheatley, Margaret J. and Myron Kellner-Roger. *A Simpler Way* (San Francisco, CA: Berrett-Koehler, 1998)

Wheeler, David A., 'More than a Gigabuck: Estimating GNU/Linux's Size', *www.dwheeler.com*, July 2002.

Available from http://www.dwheeler.com/sloc/redhat71-v1/redhat71sloc.html
Williams, Eliza, 'The Future of TV?', *Creative Review*, August 2006
Wright, Robert, *Nonzero* (Abacus, 2001)
Zeldin, Theodore, *Conversation* (Harvill Press, 1998)
Zittrain, Jonathan L., 'The Generative Internet', *Harvard Law Review* 119. 1974 (2006)
Zuboff, Shoshana, and James Maxim, *The Support Economy: Why Corporations Are Failing Individuals and the Next Episode of Capitalism* (Allen Lane, 2002)

Web addresses

www.blizzard.com/inblizz/profile.shtml
www.bookcrossing.com
c2.com/cgi/wiki?WikiHistory
counter.li.org/
english.ohmynews.com/
www.fark.com
www.ige.com
www.plastic.com
portal.eatonweb.com
www.slashdot.org
www.technorati.com/about
www.worldofwarcraft.com

INDEX

J